LEST WE
FORGET

LEST WE FORGET

A Doctor's Experience
with
Life and Death
During
the Ebola Outbreak

Kwan Kew Lai, MD, DMD

VIVA
EDITIONS

Published in the United States by Viva Editions, an imprint of Start Midnight, LLC, 101 Hudson Street, Thirty-Seventh Floor, Suite 3705, Jersey City, NJ 07302.

Printed in the United States.
Cover design: Allyson Fields
Cover photograph: iStock
Text design: Frank Wiedemann

First Edition.
10 9 8 7 6 5 4 3 2 1

Trade paper ISBN: 978-1-63228-062-6
E-book ISBN: 978-1-63228-063-3

Library of Congress Cataloging-in-Publication Data is available on file.

Names of some people depicted in this book have been changed to protect the privacy of the individuals.

For Scott, Tim, Cara, and Charles

For all who were and are still affected by the Ebola virus and the Ebola fighters who put their lives on the line.

EBOLA CLAIMED 11,325 LIVES as it tore through West Africa between 2014 and 2016. But that number does not reveal the nightmares of the crisis. Only a relatively few number of people witnessed the days of the dying, and know what a horror it is. Infectious disease doctor Kwan Kew Lai is one. She recorded dozens of such scenes during several weeks she spent trying to save patients in Ebola wards in Liberia and Sierra Leone.

In addition to the gory details of blood, diarrhea and vomit, Kwan Kew recounts the emotional toll of Ebola. Adults and children infected with the disease often died in Ebola wards alone, with no loved ones beside them. They couldn't be hugged or stroked by doctors and nurses because of fear of infection. Furthermore, their terror was transferred to patients on neighboring cots who looked on in horror at their not uncertain fate. Kwan Kew recalls one young man in the Ebola ward, who sat bolt upright as a freshly dead corpse was unceremoniously sprayed with chlorine and bundled into a white body bag. She also recounts the mourning of children separated from their parents, and stoic or frantic mothers who watched their babies die.

During her second stint as an Ebola doctor, Kwan Kew also witnessed suffering from malnutrition, childbirth and tuberculosis. The health systems of the three countries hit hard by Ebola were among the weakest in the world, and Ebola sucked up whatever resources were available. Worse, people who had maladies other than Ebola risked being infected when they sought care at hospitals.

In sharing the personal stories of her patients' final days, Kwan Kew honors the dead. So many Ebola victims were buried hastily in shallow graves in the midst of the crisis. These accounts at least show that someone heard their cries.

Kwan Kew's tales also offer up questions that policy makers and global health experts might consider now that the crisis has subsided. For instance, months passed as international aid groups figured out how to send volunteers. And upon return to the United States, Kwan Kew and other doctors were subjected to shifting sets of security requirements. Some make sense, such as temperature monitoring, but many were absurd. After flying on crowded flights, Kwan Kew was once detained at an airport without explanation. Perhaps the US might consult epidemiologists now on the rules of re-entry, so that the country is better prepared to respond in the event of another pandemic.

Other ruminations have to do with the way patients were treated. For example, Kwan Kew repeatedly expresses her desire to give patients medications to ease their pain and anxiety—but the clinics she worked in were not prepared for this intervention. She also writes about how Ebola doctors and nurses were treated differently based on their nationality. International humanitarians received the highest care that medicine could offer when they were suspected of having Ebola. Local staff were not. It is an uncomfortable truth that Kwan Kew does not shy away from discussing.

Ebola will strike again. Indeed, smaller outbreaks have since hit the Democratic Republic of Congo multiple times. Only by remembering the past can the world learn how to make these crises less devastating.

AMY MAXMEN

Maxmen is a journalist who covered the Ebola outbreak for National Geographic, Newsweek, Al Jazeera, the Economist and other outlets. She is currently a senior reporter at Nature, based in San Francisco.

PROLOGUE

WHEN I WAS YOUNG, my father boasted to his friends that I was the smart one and one day he was going to send me to Australia to attend university. But that did not jibe with my family situation. After his mandatory retirement as a government civil servant, my father worked long hours seven days a week running a food stall in a restaurant. He had twelve children to feed, clothe, and send to school, and what he earned was barely enough to hold the family together, let alone send me to a university, local or abroad. We siblings helped him after school and on weekends. There was no such thing as a day off, much less a vacation. My resourceful mother made food appear at the dinner table, even if it was sometimes the only meal that day.

I was born and brought up on the tiny island of Penang, off the western coast of the Malay Peninsula. Wealthy families from Penang would send their children to England, India, Australia, or America for higher education, but for me graduation from secondary school spelled the end of my schooling. I might have followed in the footsteps of siblings or friends and attended a nursing school or teachers' college on government stipend, but I was not interested in either profession. The Malaysian government awarded scholarships to the University of Malaya, but gave preference to bumiputras—Malays or indigenous people—and I was no bumiputra. (My paternal grandparents were immigrants from China, and my mother is an orphan.)

One day a young acquaintance of my sister who was home from an American college for summer vacation told her that some colleges offered full scholarships to foreign students. This glimmer of hope sent my best friend and me to the United States Information Service library in Penang, where we spent innumerable hours researching such institutions. I knew that I would have to get a professional education if I was ever to escape the cycle of poverty.

On March 17, 1970, after a long search and many applications, Wellesley College offered me a full scholarship. The pristine, pastoral New England campus and impeccable homes I got to know while enrolled at Wellesley were a sharp contrast to the squalor and dilapidation I left behind. I was thrilled to be in the United States but saddened by the enormous disparity between it and my childhood home.

I wanted to go to medical school, but in the 1970s admission into medical school was fiercely competitive and my foreign student adviser steered me toward public health or dentistry. Harvard School of Dental Medicine accepted me right after my junior year at Wellesley, but once there I realized that my first love was still medicine. So, after completing my dental education, I went on to get my medical degree. Eventually I specialized in infectious diseases.

Growing up I had heard about the humanitarian work of Drs. Tom Dooley and Albert Schweitzer. I always wanted to contribute in a similar way to help people live a better life, but for many years having a family of my own took precedence. In the aftermath of the terrible earthquake and tsunami of 2004–05 that devastated communities all along the Indian Ocean, I spent several weeks as a volunteer doctor in India. Finding this experience deeply satisfying, I left my position as full-time professor of medicine. Slowly I carved out a part-time position in clinical medicine that enabled me to volunteer for several months in any given year. Over the last decade, I have contributed my medical services during humanitarian crises in more than a dozen countries, including Vietnam, Tanzania, South Africa, Nigeria, Haiti, Libya, Kenya, Uganda, South Sudan, Malawi, the Philippines, and Nepal.

MELIANDOU, A VILLAGE in Guéckédou, a remote region of Guinea, is home to some thirty families living in dilapidated stucco or mud-brick houses with tin or thatched roofs, some weighted down with rocks. It sits deep in the jungle, surrounded by low hills that are largely deforested. Tall coconut trees, oil palms, young banana shoots, and cassava bushes are interspersed with towering reeds, elephant grass, and dense undergrowth. The heat, humidity, and soaking rain are unforgiving. On December 26, 2013, a two-year-old village boy, Émile Ouamouno, came down with a mysterious illness of fever, black stool, and vomiting. Two days later he succumbed to his sickness. In the ensuing days, Émile's three-year-old sister, Philomène, his pregnant mother, Sia, and his grandmother were also struck down. Etienne, Émile's father, was left to grieve alone.

Not long after mourners attended the grandmother's

funeral, the illness quickly spread beyond the village. It marched unflinchingly into Guinea's capital, Conakry, when one of the Ouamounos' sick relatives arrived there: he became the spark that ignited the epidemic. Health care workers contracted the illness while caring for infected patients and spread it to their relatives and colleagues who in turn were looking after them.

For the next few months the chain of transmission and devastation continued unabated. The infectious agent was as yet unidentified: initially thought to be cholera, its symptoms were also similar to other infectious diseases common to the tropical region, such as Lassa fever. By early March 2014, Guinea's health officials and Médecins sans Frontières (MSF; known in English as Doctors Without Borders) and the World Health Organization (WHO) concluded that there was a common link among all cases, but did not determine what it was. On March 22, the Institut Pasteur in Paris identified the virus as Ebola Zaire, the most lethal of the five distinct species of EVD, or Ebola virus disease. In the past this virus had confined itself to such equatorial and East African countries as Congo, Gabon, Sudan, and Uganda; now it had cropped up in West Africa, some two thousand miles away. On March 23, 2014, WHO published its first outbreak report, covering forty-nine cases and twenty-nine deaths. By then the first few cases of Ebola had spilled across the border into Liberia and Sierra Leone and marched resolutely into these nations' capitals, Monrovia and Freetown, baffling and overwhelming their already weak and fragile health care systems. At the end of June 2014, MSF sounded the alarm: they were already stretched beyond their limits and asked for other international players to step into the fray. Many more health care volunteers would be required for this unprecedented outbreak. WHO called a regional meeting for early July in Ghana.

Through a retrospective investigation conducted by WHO, it was discovered that Émile (or Patient Zero, as he came to be called) had been known to play near a hollow tree infested with fruit bats, known carriers of the Ebola virus. The timber and mining industries have deforested vast regions of Guinea, driving fruit bats closer to human habitats. Four months after Émile died, Meliandou had buried fourteen of its residents, and by the end of the year, the death toll in Guinea, Liberia, and Sierra Leone stood at over seven thousand people.

I volunteered to help during the Ebola outbreak of 2014, because I could not see myself staying at home when there was a real need for health care personnel on the ground in West Africa to help combat the epidemic. I knew volunteering was dangerous, but I went into it with little knowledge of just how dangerous it could be. At the end of June 2014 I was on holiday with my family in Wellfleet, Cape Cod, when I received an email from Medical Teams International (MTI) urgently asking for a volunteer to be deployed almost immediately to Monrovia, Liberia, to assist in dealing with the Ebola epidemic. My first mission with MTI had been for the cholera outbreak in Haiti following the 2010 earthquake there, and I was on their emergency relief team, whose members are usually emailed whenever MTI sends teams to crisis areas. I offered my services even though it meant I might have to cut short my holiday. My response noted that although I was an infectious disease specialist with many stints as a volunteer in Africa, I had never seen a case of Ebola. For whatever reason—MTI rarely explains how or why they select their doctors and nurses for a particular assignment—they decided fairly quickly to send instead a pediatrician who had tropical disease training.

That summer news of the Ebola outbreak was all over the media. The human toll in the African countries affected was staggering and difficult to comprehend—especially the contagious aspect. During the last week of July, the first American to contract the disease was identified at a hospital outside Monrovia: Kent Brantly, a doctor working with the Christian medical mission Samaritan's Purse. Ten days later a second aid worker with the same organization, Nancy Writebol, was diagnosed, and soon both were evacuated to Emory University Hospital in Atlanta to be treated in the isolation unit there. Government officials were caught off guard and a few flickers of panic rippled over the nation's airwaves: Were America's "good Samaritans" importing the dread disease onto our pristine shores?

This kind of overreaction was aggravated by confusion on the part of many health organizations. Perhaps by chance, MTI's tropical disease specialist landed in the Paynesville City treatment unit, the same hospital in Liberia where the two Americans had come down with Ebola. Fearing for his safety, MTI immediately recalled their sole volunteer and put a temporary moratorium on sending more medical personnel.

Ebola raged on. With each passing day, the situation became more dire, and my feeling that I had to be there intensified. Watching the nightly human drama on television, I could not hold back my tears. Since MTI had no immediate plans to send volunteers, I began to send out inquiries to other humanitarian organizations, and I used my contacts at MSF, for whom I had worked in the past, to see whether they might be able to use me in their efforts. Unfortunately MSF wanted me to do another long stint of nine to twelve months in a non-hot zone; I was not willing to do that when it was precisely the Ebola hot zone that was calling me to respond.

By early August WHO had finally pronounced Ebola a public health emergency of international concern. USAID, this country's official foreign aid agency, took on a role as clearinghouse for NGOs involved in combating the contagion. On their website, a prompt appeared announcing, "Ebola Outbreak: Qualified health workers needed. Sign up here." The application form for volunteers asked for background information and skill sets. The few NGOs that responded asked me to go to their own websites to fill out essentially the same information. Days and weeks went by, and often the organization that finally responded asked for more or different information that could have been requested on the first form. In some ways it was understandable: NGOs were still taking stock of the situation and were uncertain what role they should be playing in a world-wide effort. Nonetheless, the process moved so slowly it was hard to believe the Ebola epidemic was really being treated as an international health emergency.

Many nights I lay in bed feeling helpless, willing to go but with no organization to send me. In late August I intensified my search. One of my avenues of inquiry was WHO's Global Outbreak Alert and Response Network

(GOARN). Alas, they turned out to be a nearly insurmountable bureaucracy, repeatedly requesting the same information and shuffling my CV and offer of services from one hastily assembled response group to another. In the end they replied with a form letter saying they were reorganizing their operational framework and directing me instead to contact the NGOs I had already approached such as MSF, the International Red Cross, International Medical Corps (IMC), etc.

During this time—September 18—the United Nations Security Council adopted a resolution for urgent action, and other UN bodies estimated that, to be effective, the response would have to be scaled up by at least a factor of twenty. On September 30 Thomas Duncan, a Liberian who had been visiting his immigrant partner and her family in Dallas, became the first person within the United States to be discovered suffering from Ebola and hospitalized. Ten days earlier he had traveled from Monrovia, without protection, on several airplanes, and in Dallas the emergency room of Texas Health Presbyterian Hospital initially misdiagnosed his problem and sent him back to his partner's home with some antibiotics, which are useless for a viral disease.

Fortunately I had never relied on the GOARN network but had simultaneously sent out an application to IMC, and by early September I was actively in discussion with them for my deployment to Liberia. IMC was slated to open an Ebola treatment unit (ETU) in the Suakoko district of Bong County in mid-September. Screening and interviews took a couple weeks. On October 1, the day after Thomas Duncan was diagnosed as the first case of Ebola in the United States,

I was finally offered a contract by IMC to work as a volunteer in their Bong ETU, leaving in two weeks.

I took a three-day Ebola training course offered by the Centers for Disease Control and Prevention at the Anniston, Alabama, Center for Domestic Preparedness, a facility run by the federal disaster relief agency FEMA. In my class were thirty-five other health care personnel from all over the country. We practiced putting on and taking off the full gear of personal protective equipment—and began to have some sense of what it would be like to work in an Ebola treatment center. Even in this simulated setting, we felt as though we were walking through a minefield. Our worst day was the last, when it reached 89 degrees and we were sweating buckets inside our impermeable outfits. Of course, in the humidity of Liberia, it would be much worse. On this same day, Thomas Duncan died in Dallas from Ebola.

My family and close friends in the US supported my ambition to serve in Liberia despite the obvious fear of possible infection. They also knew me well enough to know that once I made up my mind, there would be no turning back. Back in Malaysia, my family thought my past volunteering was enough, and that I was crazy to go on an even more perilous adventure.

Was I afraid of contracting Ebola? Sure—I am human like everyone else. But what about the Liberian health workers who stayed at their jobs or volunteered for new ones simply because they were the health care providers and the patients needed them? Hundreds of health care workers had contracted Ebola, and over half of them had died. I could not sit back and watch the death toll rise and not do something about it. If the process of finding an NGO to volunteer with had not been so tedious, I would have been there much earlier.

If at first I perhaps underestimated the risks of going to Africa to beat back Ebola, by the time I left my home for Liberia, it had become abundantly clear to me that this was a different sort of volunteering mission. By then a patient with Ebola had died while being treated in an American hospital, and two nurses who had taken care of him had come down with the infection. Going to Suakoko brought with it the real danger of my being infected or even dying. Life had been good and I was not ready to leave prematurely, but if it were really my time to go, then I would accept it. Did putting my life on the line in such a situation in fact constitute an act of selfishness? No question, going to Liberia to treat Ebola patients was going to induce high levels of anxiety in my family, however strongly they supported my freedom to choose. I grappled with my conscience over this issue but felt in the end that, given my medical skills, to help suffering humans superseded concern for my own family.

For over a decade I have worked with refugees and people displaced by natural disasters, wars, and conflicts who face seemingly endless and unmitigated struggles with inadequate shelter, lack of food and water, security, and health issues, and I have seen poverty where children bear the brunt of malnutrition. I was often touched and inspired by their stories: women with young children escaping imminent death, walking for weeks to a refugee camp with only the clothes on their backs, braving the dangers of disease and violence from fellow human beings; children who were alone, having lost their parents during their run for safety; and men and women who swam crocodile-infested rivers to freedom.

The Ebola outbreak happened in three countries that had just been through years of civil war, leaving them in desperate situations and vulnerable, with broken or nonexistent health care infrastructures. Without outside help, the epidemic would have created a disaster of monumental proportions, bringing tremendous suffering to innumerable people. Remembering all the courageous people I had met during my volunteering, who still had the unshrinking desire to survive despite their confrontations with seemingly insurmountable difficulties, I not only applauded their courage, I could not turn my back on them when they needed help the most.

My volunteering experiences have taught me that while I cannot change the world altogether, I can surely change the circumstances for a few of the world's people.

In playing a small part in the containment of the Ebola outbreak, I hope I have contributed to more than an expedient and temporary solution. Ebola is not just an African problem but a worldwide concern. No walls or barriers can stop the spread of the virus. Although the public fear of a pandemic has been allayed, public health officials believe that Ebola continues to be unpredictable and another outbreak is not inconceivable.

EBOLA VIRUS DISEASE (EVD) is an acute, serious, often fatal infection. Unlike the previous outbreaks in Central and East Africa, which occurred in remote villages near the tropical rain forests of Central Africa, the huge, volatile 2014–16 Ebola outbreak of West Africa involved both rural villages and urban areas, resulting in more infections and deaths than all past outbreaks combined.

Named after the Ebola River in Yambuku, this virus first appeared in two simultaneous outbreaks in 1976, in Nzara, South Sudan, and in Yambuku, Democratic Republic of Congo.

Scientists believe that fruit bats are the natural hosts of the virus. It can be introduced into humans when they come into close contact with the blood, secretions, body fluids, or organs of infected animals in the rain forests. Human-to-human transmission then occurs through direct contact with the blood, secretions, and body fluids of infected people or objects contaminated with these fluids.

In the beginning of the outbreak in West Africa, when it was not known what the cause of the illness was, health care workers frequently became infected while caring for sick individuals without personal protective equipment, which entails the wearing of gloves, hood, mask, goggles, aprons, and gowns, leaving no skin exposed. At the ETU we wore boots, as we had to step into basins of chlorine each time we entered and exited the unit. Chlorine or household bleach was effective in killing the virus and was used as the agent in the unit.

Before the outbreak, the three countries, Guinea, Sierra Leone, and Liberia, deep in the throes of this tragedy with inadequate health care infrastructure, had a ratio of one to two doctors for every hundred thousand people. The toll taken on the health care workers made the meager workforce even smaller. By the end of the first year of the outbreak, more than half of the seven hundred infected workers had died.

Traditional burial rites in West Africa called for the washing and oiling of the body of the deceased by family members, and because the corpses remained contagious, Ebola could be communicated through contact with the body fluids. This traditional ritual, along with the practice of moving bodies to be buried in a different village, helped to inflame the spread of the virus early in the outbreak and impeded efforts at containment.

Once a person was infected, it might take between two and twenty-one days for symptoms to occur. They included sudden onset of fever, fatigue, muscle pain, sore throat, and headache and might progress to more ominous symptoms—vomiting, diarrhea, rash, some degree of kidney and liver failure, and internal and external bleeding that manifested as oozing from the gums, nosebleeds, or bloody stools. The disease had a high mortality rate, ranging between 25 and 90 percent of those infected, with an average of about 50 percent. Death was often due to shock from low blood pressure caused by fluid loss, and typically followed one to two weeks after symptoms appeared.

In the tropics, EVD might mimic other infectious diseases, such as malaria or typhoid fever. During the outbreak, because blood and fluids from potentially infected patients were an extreme biohazard risk, the US Navy set up a laboratory with maximum biological containment conditions at Cuttington University in Bong County to process samples. The diagnosis of Ebola relied on the detection of viral RNA in blood or body fluids and was reported as the number of copies of virus detected per milliliter of blood. Studies showed that patients with fatal outcomes tended to have a higher number of copies of virus RNA than those who survived, which we referred to as a high Ebola titer.

At the time, in the ETU, patients could only be given supportive care with rehydration using oral or intravenous fluids. Our protocol in the ETU included treatment of specific symptoms like fever, replenishment of electrolytes lost through profuse diarrhea, and treatment of suspected infections such as malaria and typhoid fever. Such approaches improved survival. The unit did not have the ability to check for electrolyte imbalances, and it was our best guess as to how much potassium each patient should receive. There was as yet no proven treatment available for EVD during the peak of this outbreak.

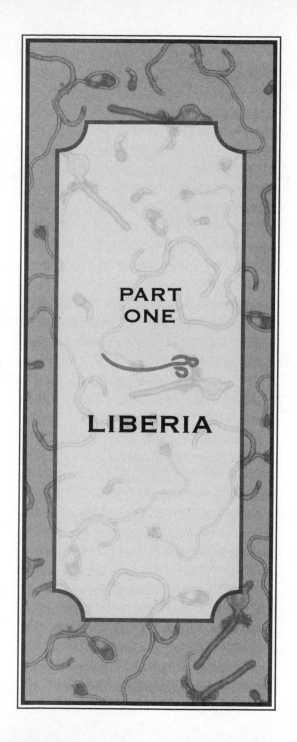

PART
ONE

LIBERIA

I

MONROVIA
TO BONG

I LEFT FOR LIBERIA on Royal Air Maroc. Most commercial airlines, including several from the United States, had ceased servicing West Africa. Brussels Airlines and Royal Air Maroc, it seemed, were the only two still flying to Liberia. I had an interminable thirteen-hour layover in Casablanca, listening to endlessly repeated piped-in music. Perhaps I should have taken a trip into the city, but since I had visited Casablanca a few years earlier and seen quite a bit of it, I did not feel compelled to play it again.

The flight from Morocco to Liberia was filled mainly with expats heading to Liberia or Sierra Leone. I wondered what was running through their minds—excitement, anxiety, fear? Upon arrival at the Monrovia airport, we had to wash our hands and have our temperatures taken. My slip of paper from the Liberian Ministry of Health enabled me to stay in the country up to six months, which meant that for once in my life I could go through immigration in an African country without having to pay a hefty visa fee. It was well past midnight. Suitcases loaded onto a conveyor belt came rolling in while porters wearing heavy-duty gloves helped to dislodge them. More than two months after my first attempt at volunteering to serve as a medical doctor, I had finally stepped into Ebola land.

A logistician and a driver from International Medical Corps picked me up. The driver dragged my suitcase over the dirt, lifting it when we passed over a mud puddle. I wondered whether Ebola was lurking in its dark water. We drove through the dark; our headlights illuminated tall reeds on both sides of the road. Monrovia was under curfew, and only essential vehicles were allowed to be on the roads. I did not get into Monrovia until four in the morning.

I slept late and had lunch around noon. That afternoon I met the IMC finance officer, an Ethiopian named Fikadu, who was slated to accompany me on the trip to Suakoko. As our driver maneuvered the four-wheel drive cruiser along the streets of Monrovia, I saw a huge sign declaring: "Ebola is real and it is here in Liberia." Markets with their colorful, sun-bleached umbrellas fluttering were in full swing, and we stopped by the roadside so the driver could buy some bread. Outside the city we were halted twice at roadblocks with improvised Ebola stations, where we had our temperatures taken. Both stops sported this banner:

Fight the Ebola virus
Protect yourself
Protect your family
Protect your community

As in most African countries, once we left the city the smooth tarmac road turned first into a potholed highway before changing into a surface of hard-packed red earth. Along the highway, life seemed to go on despite the epidemic. Dark clouds threatened intermittently, bringing short spells of heavy rain. It was the tail end of rainy season.

Suakoko is two and a half hours north of Monrovia. The evening sky was darkening as we turned onto a red-earthed dirt road at a towering sign announcing the "Ebola Treatment Unit of Bong County/Funded by Save the Children/Managed by IMC and USAID." Next to it a smaller, unassuming sign worn by rust and sun indicated the still-active leper colony of Suakoko. We drove a few minutes over an undulating, muddy red road flanked by dense groves of rubber trees whose trunks were covered with patches of green-gray moss. We then turned sharply at another ETU sign with an arrow pointing right. A cluster of blue-tarpaulin buildings loomed ahead, atop a small hill surrounded by green jungle. Ominous, dark rain clouds formed the background.

Our vehicle stopped at the gate. A man in black rubber boots and a blue protective suit and wearing a plastic face shield and a mask started to spray the lower parts of our conveyance with chlorine from the heavy tank strapped to his back. Once he had finished with the tires, we drove into a compound fenced with orange netting. The ground cover was rough gravel. Ahead of me I gazed into another fenced compound that seemed a beehive of activity: many people in scrubs and some in street clothes and black rubber boots, walking to and fro purposefully, as if each performed a well-rehearsed part onstage. Backlit by the waning light of evening, the human hubbub produced a steady stream of murmuring voices.

I opened the door of the vehicle and took a look at the gravel outside. I had been told nothing about where the green zone ended and the red zone began. I was unsure what to do, but I was not about to step out with only the slippers I had on my feet. Suddenly the reality hit me: Ebola virus could be anywhere and everywhere. Fikadu scrambled over the seat to the back of the cruiser and retrieved a pair of multicolored fur-lined polka-dot boots that looked rather feminine and exchanged his shoes for them. He then stepped down onto the gravel. He had come prepared, but my boots were in my suitcase, buried under a heavy pile of boxes and other bags. It would be a huge effort to fish them out. I called over Fikadu and explained my dilemma; he took a quick look at my feet and slippers and said I should just stay in the vehicle.

While I hesitated, I heard someone call out my name loud and clear. A physician in scrubs and black boots approached the cruiser and said that he had heard my interview on National Public Radio. It was the first I heard that the interview I had given an NPR reporter while I was being trained in Alabama had been aired the day I left Boston. In it, I had raised some questions about the difficult process of signing on as a volunteer in the Ebola outbreak. So: my reputation had preceded my arrival, for good or bad.

I watched from the vehicle as droves of people stepped into tubs of chlorine at the gate, washed their hands at the chlorine tap, rinsed them in water, and walked away to board their bus for home. I heard my name called again, this time by a woman with short black hair pulled back into a small ponytail. I liked the assured way she used her booming voice and pronounced my name without faltering. This was Som, Fikadu told me: one of the principal coordinators. It was shift-changing time, and apparently Som and the others had no time to brief me. Fikadu finished whatever he had come to do and prepared to pull out of the ETU, but first our vehicle had to be sprayed again. Not far away, in the lingering pink light of the setting sun, a crew in their personal protective equipment was spraying down an ambulance that had just delivered a sick patient. The people shrouded in PPE looked surreal. I noted that the ambulance was using a separate route leading to a different entrance, and alongside it ran a chain-link fence. Beyond us, wriggles of lightning pierced the dark clouds, and thunder began to grumble.

The ETU was located about three miles from the main road and surrounded by a dense grove of rubber trees. Another three miles or so down this road, in an area tarmacked not so long ago by the Chinese, was Cuttington University, where both nationals and foreigners were lodged. As we emerged from the deep jungle, the sky brightened. People along the road hugged the tall elephant grass on both sides to avoid being hit. There were no sidewalks, only narrow, hardened dirt paths created by the constant pounding of footsteps. We turned into the arch marking the entrance to Cuttington; the gatekeeper raised the bar to let the cruiser through. As we lumbered uphill over a deeply rutted red dirt road, several brick buildings came into view among overgrown fields of tall grass. Apparently the grounds had been left to run their course after the university closed because of the Ebola outbreak. I was dropped off at a dormitory. When I asked Fikadu if he knew when I was to report in the morning, he told me I should ask the other expats who were staying with me.

He was wrong. I did not encounter any expats that evening. I was hungry and the dormitory did not serve food. Food was served only in the ETU, I learned; someone should have told me to eat my dinner there. I rummaged for any remnants of food in my suitcase and backpack—a granola bar, a piece of cheese, a few pieces of chocolate—and washed those down with water from my Nalgene bottle. A Liberian logistics assistant named Charleslyn, who also lived in the dormitory, brought a fan and surge protector for my room, introduced herself, and explained how things worked. When I mentioned my hunger, she kindly suggested I have one of the drivers take me to the nearest town, Gbarnga, to eat in a restaurant. Gbarnga was twenty minutes away and I was too exhausted to attempt that; besides, I had no Liberian dollars. Before I retired Charleslyn came back with a plateful of rice and peppery okra, having sent a driver to get it from the ETU. The okra was too spicy and burned my tongue, and by then my hunger had largely left me.

I was to live with the nationals—Liberians who had jobs at the ETU—in Rally Hall Dormitory. My own room came with three bunk beds; the two unused ones had been pushed into a corner. The shelves in all the closets save one were broken. The bed on the lower bunk of one of the beds tucked against the walls was made; gossamer mosquito nets hung from the top bunk. Besides the two empty bunk beds the only furniture was a student chair. A dim, naked lightbulb dangled from the ceiling, making the room even gloomier.

There was no running water in the bathroom except for a few showers that worked, and the toilets did not flush. In a corner stood two big blue barrels with potable water. Only two of the toilets had doors—which the girls would never close—and few of the showers were fitted with curtains (probably placed recently by the IMC). The girls were also used to showering without curtains. Modesty was not a habit of the nationals, and they often walked topless even in the presence of men. So much exposed skin in public left me speechless more than once.

Luckily, I really did not have any kind of expectations about the place I would be staying. By the time I got to Liberia, I had been volunteering in Africa for almost a decade, so my current conditions didn't shock me. I had been to more desperate places, where I had used pit latrines and taken bucket baths in sheds with rickety doors.

I dusted off the single shelf in the closet and unpacked. Charleslyn had advised me to padlock the closet, but the hinges that would hold the padlocks were flimsy and loose and could easily be pried off. I showered under some trickling, lukewarm water coaxed out of the showerhead. Then, exhausted from the long day, I crawled into the lower bunk bed and put on my headlamp to read myself to sleep. Work would start at eight in the morning, according to Charleslyn.

II

DIARY: "THESE ARE HARD TIMES"

October 17, 2014: Suakoko, First Day at the ETU
The tall ceilings and transoms of the dormitory echoed back all the noises within, magnifying them and making sleep an impossibility. Before the morning light graced the eastern sky, a heavy rainstorm tumbled from the heavens. Cool air wafted through the open windows from the courtyard. I stumbled into the bathroom and used cold water bailed from the large barrels of water lining a wall to wash my face and flush the toilet. A few of the sinks lacked drainpipes; water flowed straight out of the basins into a bucket below. By evening the buckets were full and water overflowed to the floor. To brush my teeth, I was careful to use only the water from my Nalgene bottle. So much of the building's infrastructure needed fixing it made my head spin. How had such a well-known private university been allowed to sink to this sad state of affairs?

A robust voice calling my name jolted me back to reality. I turned and faced Som, the ETU program director who had called out to me yesterday while she was walking the grounds of the treatment center. She told me that the medical staff began work at seven A.M., not at eight, as Charleslyn did. If I could be ready in fifteen minutes, Som said I could leave with her at six forty-five. Otherwise I would have to wait until seven for another ride to the ETU.

It took about fifteen minutes to reach the Bong ETU; the sky was brightening quickly as we rode along the highway.

At the morning handover, the night shift informed us of one death. There were thirty-three patients, fifteen with confirmed Ebola and eighteen suspected. The ETU was capable of holding fifty-two beds and had opened only weeks before, in mid-September. Etiquette in Liberia during this outbreak allowed for no touching, not even a friendly handshake, so I had to consciously refrain from shaking hands with any of my fellow volunteers.

The enclosed area where our cruiser had parked the night before was actually in the green zone, a safe area where the administrative offices were located. (So I *could* have stepped down from the car last night, even in my slippers.) Just outside this area on the roof of a storage room sat two gigantic vats of chlorine. Beyond this green zone was the intermediate zone, home to the nursing office, handover room, pharmacy, changing room, the doctors' and psychosocial teams' offices, and the WASH (water, sanitation and hygiene) room. Outside the WASH room were lines for drying aprons and stakes in the ground for drying rubber boots. Inside, the washing machine and dryer were constantly churning with the sounds of scrubs being cleaned. Clean goggles and heavy-duty rubber gloves were hung to dry from the nails in the rafters.

Orange net plastic fences separated the different zones. Beyond the intermediate zone was the red, or hot zone, where the triaging area and the suspected and confirmed wards were located. To avoid possible contamination, one was allowed to walk only from the suspected to the confirmed ward, not the other way around. Two showers stood guard between the intermediate and hot zones, home to the donning and doffing areas. The triaging area was where ambulances dropped off patients suspected of having Ebola. In the triaging room, which was extremely hot and humid, patients sat more than six feet away from the interviewers, separated from them by a wire-mesh screen. The interviewers wore masks, face shields, and gloves. Once patients had been escorted to the suspected ward, sprayers with tanks of 0.5 percent chlorine solution copiously sprayed first the waiting area, then the paths just taken by those patients.

The suspected and confirmed wards were two long, blue-tarped pavilions with tin roofs, gutters, and a row of high windows cut into the tarp along both sides. Each ward currently held twenty-one beds. The suspected ward had only single rooms, so as to prevent the spread of Ebola to patients who would later test negative. The confirmed ward had rooms with one, two, or three beds. Outside each ward were booths where callers could sit and visit with patients, again at a safe distance and separated by a wire-mesh partition that kept them from touching the patients.

The patients' showers and toilets were at the back of the wards. A little way from the confirmed ward stood the morgue. It was visited frequently by raucous crows. Down the slope on the western side of the ETU was the incinerator, where contaminated trash was burned at all hours of the day, throwing off a brown smoke that trailed slowly into the sky.

My main impression of the morning was a lot of hand washing with 0.5 percent chlorine and dipping our boots into footbaths as we moved from one place to another. I was rearing to get inside the wards, but Som insisted that, for safety's sake, I should confine myself on this first day to repeating the donning and doffing process. So many things happened early on, however, that we did not get to this exercise until the hottest part of the day. There were only extra-large sizes of the more impermeable yellow Tychem suit, so I wore the more breathable white Tyvek personal protective equipment. Even for that, the smallest size was medium, and I had to fold the suit around my waist to shorten it. For the ETU we wore three pairs of gloves, the first and third pairs surgical and the second of nitrile latex, securely duct-taped to our suits. The donning sequence ran like this: first a pair of surgical gloves, then nonsterile gloves, space suit, N-95 respirator, hood, apron, goggles, and last the third pair of gloves duct-taped around the edges where glove met gown. (The N-95 respirator or mask is particularly important, since it is finer than a normal surgical mask and screens out very small viral particles as well as bacteria.) This dressing process could take fifteen to twenty minutes, after which a couple of monitors in the changing room checked to ensure that no part of our skin was exposed. Last and most important, the monitors wrote our names on the hood over our foreheads. Without this name we would be hard to tell apart.

The doffing procedure started with the sprayer spraying our suit and gloves with 0.5 percent chlorine, and each step was interrupted by hand washing with chlorine, including of gloved hands. First the duct tape was removed and off came the third pair of gloves, then apron, goggles, hood, suit, N-95 mask, and the last two pairs of gloves. The final step was hand washing in 0.05 percent chlorine and then rinsing with water.

In the heat of the afternoon, we walked away from the ETU to a forested area where the cemetery was located. Four men in PPE had just returned from there with an empty stretcher. When we arrived, a group of men, skin glistening with sweat, pants coated with dirt, were digging fresh graves for the newly deceased, including the patient who died last night. Chopped ends of sacrificed saplings protruded rudely from the mounds of freshly dug mustard-colored earth. Surrounding the cemetery were dense growths of trees with hanging dead branches and lifeless tendrils. A relative stood several feet away, watching silently. Apparently, many relatives chose not to come to the ETU to say their last goodbyes, whether out of fear of infection or having to travel a long distance. Beyond the grave diggers were a dozen or so grave markers, white plaques attached to wooden stakes driven into the ground with these words inscribed on them: In Loving Memory of so and so, sunrise (date of birth), sunset (date of death). These were the graves of the first group of patients to have lost their lives in the ETU. One of the grave plaques indicated a baby who had died thirteen days short of his first birthday. Nearby, several members of a family were buried close to one another. There was no breeze, the air was stifling; one could hear a pin drop in the deafening silence. This place marked the untimely end of someone's journey on this earth, life snatched from them before they had reached their full potential. Surveying this bleak landscape, a strong emotional upheaval stirred deep inside me, reminding me that life was fragile, ephemeral, and fleeting.

There had been close to thirty deaths since the ETU opened on September 15.

Closer to evening we got our patients' results on the Ebola polymerase chain reaction (PCR) test. Before the US Navy came to set up the Ebola laboratory at Bong, it took four days for test results to arrive from Monrovia; now it took four to six hours. Gloria, who had been in the confirmed ward of the ETU for twenty days now, was still waiting to be clear of the virus. Several patients in the suspected ward had tested negative, but it was too late to discharge most of them that day. When we delivered the good news to the patients who tested negative, they were visibly relieved; a few clasped their hands in supplication, praising God for sparing them.

A few patients who tested positive needed to be moved, so we gowned up in PPE. It was harder for us to break the bad news to the ones with Ebola, although some seemed resigned to their fate and remained disquietingly calm. We tried to reassure them that this did not mean they would not recover and softened the blow by telling them they would do well if they stuck to the treatment. After they packed their meager belongings into the two buckets that each patient was allotted when they entered the ETU, we rounded them up. In the fading light of dusk, we grimly escorted them, walking in single mirthless file along the net fence, to join the ranks of the Ebola infected in the confirmed ward. I often wondered what raced through their minds. Were they thinking that they were inevitably plunging head-long toward death? Did they wonder if they would see their families again?

The goggles given me were so huge that by this time they had begun to pinch my nostrils, making breathing difficult. Soon the lenses fogged up and I stared at the macabre surroundings through a thick haze, as if I were underwater, snorkeling.

In the midst of this flurry of activity, a recently arrived patient in triage met the criteria for suspected Ebola, so I went with another nurse to fetch him. Either he was too weak to get up to walk or he was absolutely petrified at the sight of all of us in space suits. Outside, rain clouds hung low in the sky. He sat, seemingly frozen in his chair, staring blankly at what must have looked like hovering ghosts. Eventually we were able to help him into the suspected ward. His pants were soiled with loose stool, so we changed and cleaned him. He was breathing fast and looked dehydrated, and there was a definite glint of fear in his eyes. We had no oxygen to ease his laborious breathing, but we gave him a bottle of oral rehydration solution (ORS) and instructed him to drink. The night shift would start an IV. I wondered what it was like to be cared for by strangers dressed in hazmat suits without the comfort of human touch.

As we departed, the rain came pelting down loudly on the tin roof, offering little solace to this lonely man covered by a white sheet, left in a strange, spartan room lit by an energy-saving spiral bulb. In my heart I feared the worst for him.

October 18, 2014: A Lonely Death in the ETU

When we climbed into the cruiser this morning, a few of the expats were carrying small chlorine sprayers with them. Having yet to receive a full briefing, I did not know what could be brought in and out of the ETU, and I wondered about these sprayers. Perhaps noting my curiosity, someone shoved a half-full one at me. I asked what it was for. She told me it was full of chlorine solution from the ETU and that I should bring it back to the dormitory tonight and spray the shower area or anything else I thought might be contaminated.

As we drove up the final slope to the treatment center, the chain-link gate with its heavy plastic banner— "Bong County/Ebola Treatment Unit"—loomed into view. Piles of dirt lay to one side of the road, evidence of the recent construction. In the east, the pink and orange of the rising sun mingled with the purple and blue of the sky.

The ETU that had risen in this jungle now sheltered hopes but also laid bare our human vulnerability. This morning there were twenty-six patients, fifteen in the confirmed ward and eleven in suspected. Many would be discharged today, since they had tested negative yesterday. The local nursing staff worked a six-hour shift as required by Liberian rule, while the expats worked for twelve-hour stretches. A passenger bus discharged workers for the morning shift who, after coming off the bus, picked up their boots and changed into scrubs.

Today for the first time I would do rounds in both the suspected and confirmed wards.

After morning handover, the expats met in the doctor's office to give updates on the various clinical, administrative, and logistical issues: WASH, transport, finance. Across from us in the nursing office, where the nationals met before the start of their day, came rousing choruses of song, hallelujahs, clapping, and praying. The loud noises distracted me: I wanted to go join in the camaraderie. While the expats ended their meeting with the cheer of "Go and save lives," the nationals asked God for protection for themselves and their patients. I also yearned for the cloak of God's protection.

Around eight in the morning, the cleaners and sprayers were the first to head into the wards to empty trash and spray. Red bags of hazardous waste were trundled down the gravel paths to the incinerator. The cleaners and sprayers were followed closely by the nurse's aides, who brought in the breakfasts, clean clothes, and sheets. The nurses, physician's assistants, and doctors followed. We were to drink at least half a liter of ORS before entering the ETU, but I found it both sweet and salty, and unpalatable. Forcing myself to down half a liter was a tall order.

This morning, to my horror, I discovered that the expats were served only two meals a day. For me this meant no breakfast two days in a row. There was no food or water stocked in the dormitory, and yesterday after I arrived at the ETU, I had started work right away. I could not conceive of going on rounds inside the ETU hungry. Drinking ORS on an empty stomach made me nauseous, and I was afraid that I might faint without nourishment. The nationals, I learned, were given vouchers for meals and served breakfast in the nursing office; rounds began after they ate. The expats who came before us, however, lived in guesthouses with a kitchen stocked with bread, peanut butter, digestive biscuits, tea, coffee, sugar, and water as well as snacks they had bought a month ago, in Gbarnga, shortly after they arrived. The expats had complained about the breakfast served in the ETU: no bread, butter, eggs, fruits, or jams, and instead often spaghetti with tomato sauce, rice, or fried potatoes—hardly the breakfast they were used to. So they made the decision to skip the breakfast, paying only for lunch and dinner. That was why we only had two meals a day.

Doing rounds in the ETU without breakfast seemed risky to me. This morning I came out at eleven after a couple of hours and was running on nothing but fumes. Lunch was not going to be served for another two hours. After some back-and-forth with the coordinator, who was not a health care worker and didn't round in the ETU, she announced that any expat could choose either breakfast or dinner in addition to lunch, which would always be served. That was not an option for me. Where was I supposed to get dinner? She suggested making do with crackers and peanut butter for dinner, though the dormitory had none. Or I could make a trip to Gbarnga. But when would I have time off to do that? I was willing to pay extra to get three meals a day for the sake of my health and sanity, and by the end of the discussion, the logistician agreed to arrange for breakfast at the ETU. (Surprisingly, from that day forward, all incoming expats agreed to the extra payment and ate all three meals on-site.)

In the wards the energy-saving spiral lightbulbs burned 24-7. Despite this concerted effort to save energy, the electricity was not switched off during the blindingly bright tropical days, when the light through the windows was plentiful. Indeed, I could not find a single light switch; the electricity must have been centrally controlled. The only way to switch off a light was to loosen the bulb from its socket. Many patients slept with the light glaring down from the ceiling above them.

John, the petrified patient whom we had helped into the suspected ward at the end of the day yesterday, had lived through the night but was still struggling to breathe. His IV line had stopped working. I made him a straw out of some IV tubing and sat him up so he could sip his ORS before a new line could be placed. He eagerly drank about ten ounces of fluid in between gasps but eventually slumped down in bed, exhausted. I stuck around as much as I could, rubbed his back, and squeezed his shoulders to convey the message that he was not alone. I have always felt the need for human touch when I am sick. In the US on more than one occasion I have hugged my most distraught patients. How profound must have been this man's loneliness, how tremendous his fear as he fought for his life in an alien place with no family members by his side! I wished we had morphine or IV Valium and could have eased his labored breathing and anxiety, but there were no such medications in the pharmacy in the ETU. It was only stocked with antibiotics, antimalarial medications, and paracetamol (acetaminophen), an antipyretic that was also the only pain reliever available to the patients.

We moved over to the confirmed ward. My scrubs were soaked with sweat. To ensure our safety in the ETU, each of us was paired with a buddy. I gazed up at my partner. Sweat was literally pouring down his N-95 mask, sleeves, and gloves, and he looked faint and dazed. Several times I asked if he was all right and suggested we could make our exit. No, he insisted: he still had a few minutes left in him. The heat was stultifying, and soon sluggishness stole its way into our studied movements. I read somewhere that health workers lose a pound and a half of body weight— all fluid—after an hour or two inside the wards. We stayed inside for close to an hour and half. In the end my partner was so exhausted he declared defeat: he had been bested by a petite Malaysian.

In the afternoon, we learned that John had passed away. His Ebola sample had been missed by the morning pickup and was still sitting in the fridge. He was forty-two years old.

Some recovering patients were sitting outside as the sun set in hues of yellow, orange, red, blue, and purple. *John will never have another chance to see this scene*, I thought: the life of the first patient I had helped into the suspected ward of the ETU had been snuffed out. In his dying hours, he was all alone.

October 20, 2014: Night Shift

Walking into the first room of the confirmed ward for the first time two days ago, we found eight-year-old Ryan lying on his side, neck extended and face uplifted. My partner Mike felt a pulse, but it was only his own heartbeat. Ryan's chest remained still; no brachial pulse.

The lighting was not good enough to see his pupils. My first instinct was to shield Solomon, the youngster lying in bed across from him; he had been staring at his cousin for heaven knows how long, witnessing death. On the hard cement floor was Ryan's six-year-old sister, Christine, curled up sleeping. She usually shared the bed with her brother. Had she perhaps crawled out of bed and gone to sleep on the floor because she was distressed by her brother's last struggle?

When Mike picked her up, her pants were soaking wet with watery stool. If there were a time when children needed their parents most, this was it. But Ebola is unforgiving: no one can wear PPE for extended periods of time in this heat and humidity, and without PPE, the parents would also have become infected.

I was asked to do the night shift. The last time I had done an overnight shift was during my residency years, decades ago. I joined Kent, an emergency room doctor from California; Peris, a nurse from Kenya; our entourage of local nurses; and a physician assistant. At handover, the day shift reported only two suspected and fifteen confirmed patients. Many of the confirmed, however, were not doing well and would require hydration with intravenous fluids.

One happy piece of news: Gloria's Ebola test was finally negative after twenty-three days. She beat Ebola and would return home in the morning triumphant, one of the just over 50 percent of people who survived the deadly virus.

The night WASH staff stood outside the administrative office silhouetted by the dark, threatening rain clouds, and as usual started their shift with loud prayers and hallelujahs sung from bursting chests. Once again, we expats remained quiet.

Even in the cooler evening, putting on our PPE brought on a profuse sweat. My goggles fogged up. I reapplied the defogger to them and asked the monitor to allow me to let them air-dry a few minutes before launching into the suspected ward. Kent was concerned about my being swallowed up by the medium-size outfit and afraid I might trip over my overly long apron, which grazed the top of my boots.

As we walked into the suspected ward, we were confronted by an unresponsive elderly woman who showed weakness on her left side; we were unsure whether it was an old stroke or a new problem. She had been brought in with symptoms of Ebola and died a few hours later, in the early morning.

There inside the ward, chlorine fumes filled the air. Not only did the sprayer clean the ward as we made our rounds, but big vats of 0.5 percent chlorine with taps were scattered along the halls of the wards. They sat on wooden platforms at waist level, and in between patients, we turned on the tap to wash our gloved hands. Buckets on the floor caught the chlorine.

Last night as we began rounds, big flying insects fell into the open buckets of chlorine and struggled to get out. I felt I should rescue them but hesitated to pick them up with my gloved hands, so I tried to pick them up with whatever I could lay my hands on. Inevitably they struggled off my various life preservers and ended up back in the bucket. Finally I reached in with my gloved hand and scooped them out. Of course I knew the bucket was filled with contaminated fluid, so afterward I washed my hands copiously with chlorine.

They were not the only insects. Gigantic moths fluttered on the chlorine-soaked floors, desperately trying to fly off with their wet wings. Giant rhinoceros beetles with menacing horns, seemingly drawn to the big basins of chlorine solution placed at the entrance and exit of each building for us to clean our boots in, crawled around the footbaths. In doing so they risked being crushed by the stomping boots of the health care workers. It was night, so they flew into the air in droning vertical flight, then suddenly dive-bombed, kamikaze fashion, onto any unfortunate soul, sometimes biting.

In the confirmed ward, there was a flurry of activity—the hanging of IV fluids and words of encouragement and exhortation to the patients to replace their loss of fluids by drinking the bottles of ORS we offered. Fever and watery stool in an eleven-month-old baby, Beyan, prompted more IV fluid for him. He was listless but luckier than the other children in the ward—Christine and her cousin Solomon—in that he had family with him: his father, Papa George, who had recovered from Ebola and come into the ETU to care for his son.

Since witnessing the death of his cousin two days ago, Solomon had remained wide-eyed and silent, avoiding eye contact, refusing to drink or eat; foam boxes of food were piled at the end of his bed. When we came by, Christine was sleeping curled up in the next bed and, like Solomon, showed no interest in her food or ORS. The two were moved out of the room where Ryan, her brother, had died, and she had an entire adult bed of her own to sleep in, though it dwarfed her small body.

A nurse hovered lovingly over her, singing a lullaby. Recently, Christine had also lost her mother to Ebola.

Heavy rain came pelting down on the tin roof, and it was impossible to hear ourselves speak. We moved methodically from room to room, attempting to empty IV bags by squeezing them manually or using blood pressure cuffs.

It was nearing midnight when we finished our rounds. After doffing, we retreated to the doctors' office in the safe zone. The lightbulb here was attracting too many insects, so we unscrewed it. Outside the screened window another light burned, beckoning hundreds of different species of insects in swarms, including some enormous praying mantises. Our artificial lighting rudely penetrated the profound darkness of the jungle. Instead of clarity we were probably bringing chaos and confusion to this impenetrable forest of exotic creatures. We were the aliens, invading their dark world, perhaps spelling their demise.

We tried to catnap while sitting on the hard plastic office chairs, waiting for the second set of rounds. Some nationals lay sprawled on mattresses they had stashed in the nearby changing room, clutching blankets taken from the storeroom that were meant for the patients. Peris rudely snatched these blankets away, unafraid to wake them up; she was incensed, since some patients were sleeping without blankets. Surprisingly, the nights could be chilly, at least for some Africans.

We made our second set of rounds at two thirty A.M. Many patients had diarrhea and were shivering from wet and cold, too weak to help themselves. Many clutched at their bedsheets with that frightened deer-in-the-headlights look. Cleaning and changing them all was quite a feat.

Another round of fluids and medications. Despite the cooler night, I still sweat heavily through my scrubs during my hour and half inside the wards for each set of rounds. Tucked in the last room on the right side of the ward was Varney, writhing in bed, wiry and thin, fidgeting and anxious. Even in the dim light, his bleeding gums glistened menacingly, and the old blood in the corners of his mouth had congealed and crusted. His salt-and-pepper hair glinted; he looked like he could be in his late fifties. An IV bag hung from a big, rusty nail on a wooden post of the wall, the pearl-like droplets of fluid dripping slowly. Though the evening report had said he was semiconscious, he could be awakened.

Two instances during doffing caused me some concern. The first came after I had removed my goggles and was peeling off one layer of my gloves, which had been sprayed with chlorine. I felt drops of fluid—chlorine or contaminated fluid mingled with chlorine—fly into my eye. The second occurred when the sprayer sprayed chlorine onto my Tyvek suit and some wetness seeped into my scrubs. True breaches or not, I had to live with the resulting anxiety—waiting for a full twenty-one days, the incubating period of the virus, to pass.

Dawn came. The eastern sky brightened. All of a sudden, the intense adrenaline rush evaporated and my energy level plummeted. I was ready to go home and go to bed, but there were still handover rounds and administrative meetings before the shift was over. I doused my face liberally with cold water from the tap outside the donning area to keep myself awake even as I felt physically and mentally drained. The twelve-hour shift actually turned out to be more like fourteen after handover rounds and meetings. Once again I had the urge to join the soulful singing of the nationals, longing for the protection of their collective prayer wafting skyward to heaven.

October 21, 2014: Shattered Lives in the Battleground of Ebola

The only remaining patient in the suspected ward was moved into the confirmed ward after her Ebola test came back positive. She was the daughter of a pastor who had died of Ebola a few weeks ago, before I came to Bong. In the late afternoon, the usual time for the ambulances to bring in patients to be triaged, fifteen more patients were admitted. There were two families. One was a woman with her two children. The others were family members of Simon, a patient who had been in the confirmed ward for a few days now, gravely ill. They included Simon's wife, their baby, and his father.

Elvis, the head of the ambulance team, had been making daily rounds of the villages of family members of patients in the ETU. This tracing of contacts was one of the reasons there had been such a big number of admissions in the last few days.

Before we got organized for our evening shift, the ETU was plunged into total darkness. The generator had stopped working. We searched for flashlights; I remembered seeing a stash of them along with some batteries in a box underneath a counter in the doctors' office. After a bit of loading batteries into the flashlights, I left the office and took a rare moment to gaze at the starry sky before power was restored. More stars appeared as my eyes adjusted to the darkness, but alas, I was disappointed: the sky was cloudy.

In the suspected ward, the woman who had come in with her two children was not very ill, but she looked worn-out. Winner, the younger and sicker of the children, snuggled against her while her older sister, wide-eyed and quiet, sat on a mattress on the floor next to her mother's bed. Winner cried with fright when we approached, clinging to her mother more tightly; she was weak and needed intravenous hydration. Because she kept pulling her arm away whenever a nurse tried to insert a catheter, it took several attempts before a line could be placed.

The first few times I'd met Simon, he was either sitting ramrod straight on the edge of his bed, eyes tightly shut and lips sealed, or he was on the bucket commode with legs shaking, struggling with diarrhea but still very much upright. Simon was tall and lanky and kept himself vigorously hydrated. In the wee hours of the morning, however, he was in extremis, eyes and lips still tightly shut, his breathing more labored and his neck stiff as a board. He drank eagerly when offered ORS. We tucked him in, praying that his exit would be easy. His eyes remained closed and he did not utter a word. At daybreak one of the health workers asked a patient on the other side of the orange net fence about Simon; the patient thought that Simon had passed away. He was thirty-eight and did not even have a moment to say goodbye to his loved ones; they were still in the suspected ward.

Varney had been in a tenuous state, but he hung on

tonight, eagerly taking a drink of ORS and then juice when I offered both to him. He continued to bleed from his gums and suffered from bloody diarrhea.

Fear ran rampant inside the two wards. Each time there was a death in the ETU, the patients who were recovering from Ebola made a frantic dash to move as far away as possible from the dead person. As a result, patient locations changed from day to day. Solomon was finally eating, but Christine was interested only in drink. The staff had tried to recruit one of the older recovering ladies to care for the two children, but they would not volunteer, so Solomon and Christine struggled alone in their room. I heard from the staff that before I came, a mother deposited her nine-year-old in the ETU and did not return for her even after the girl had recovered: the stigma of having contracted Ebola still marked the child. Ironically, her name was Blessing.

In the early-morning hours, heavy rain sounded like galloping horses on the roof of the doctors' office in the safe zone. Earlier in the evening, the ER doctor, Kent, had joked that I was so small I could probably use one of the round tables in the office as a bed. The pelting rain did not wake him up and he snored softly in a corner with his head against a cabinet. Meanwhile Peris, the Kenyan nurse, covered in surgical scrubs from head to toe, was also sound asleep.

Another night in the ETU, tired and weary from our labor. At the end of my second night shift, we had thirty patients, evenly divided between the two wards.

At the far western end of the ETU, light-brown smoke rose gently from the incinerator that burned infectious wastes, mingling with the morning mist as dawn broke. It reminded me of a scene from the aftermath of a battle.

October 22, 2014: Death as a Fallen Autumn Leaf
At this point, the Bong ETU was the major training center for Ebola treatment. It was also a training ground for many health care workers from different humanitarian organizations who came for a few days of hands-on experience. Almost all the volunteer health care personnel stayed for a period of four weeks; I would be the first physician to stay for six. To accommodate my being involved in the training of transients, I was switched from night shift to a so-called swing shift, from nine A.M. to nine P.M. Missing morning sign-out put me out of kilter for a little bit, but I knew I would adjust.

Daily, death visited the ETU with unfettered tenacity. Simon was buried yesterday—his baby tested negative and was sent home with relatives, but his wife remained in the suspected ward, awaiting her result. The staff reported that she was devastated by the loss of her husband. Varney and Momoh, the two other critically ill patients when I was last on night shift, also passed on.

Winner, the seven-year-old girl, tested positive. She had been vomiting and stooling and had to be separated from her mother and sister to be cared for by Bendu, a recovering Ebola patient in the ETU. It must have been very difficult for Winner, who on the night of admission was clinging tightly to her mother. Femata, another recovering Ebola patient, eventually took on the responsibility for the care of Christine and Solomon. Three members of a family of four that had come two nights ago into the suspected ward were discharged with negative Ebola tests.

During my training as an infectious disease doctor, I read about viral hemorrhagic fevers: Ebola, Marburg, and Lassa, Ebola being the deadliest. I thought that since outbreaks of these viruses occurred in Africa, it was highly unlikely I would ever see a case of Ebola infection in my lifetime. Well, I was dead wrong. Here in the ETU, I had a firsthand look at how devastating the disease can be.

A torrential rain came down during my ETU rounds. Most of the health care workers smelled the impending deluge and made an early exit, afraid they might get stuck in the ETU during a prolonged storm. For some reason my partner failed to alert me that we too should leave, and at some point I was left alone in the ward, which was against protocol. So there I was, unaware I was alone, suddenly noticing a lot of commotion out of the WASH office: a group of people were yelling at me to come out. Not realizing my partner had left, I searched from room to room to locate him. All the while the WASH personnel were shouting, "Come out, come out!"

Finally I realized my situation. Tentatively I walked out of the sheltered area into the driving rain. Immediately it dawned on me that I would be drenched walking to the doffing area. I thought it was not such a good idea to be in the doffing area with exposed skin while the wind and rain whipped through it and likely blew contaminants all over me. So I retreated back inside, hoping the rain would soon stop. But the staff was insistent—there was a great risk to me if I got stuck in my PPE in the high humidity with no partner. So I ended up in the doffing area being showered by chlorine washes and God knows whatever else!

When I left Boston, the autumn leaves were just turning. With each rain and wind, more leaves spiraled slowly to the ground. For me the very sick Ebola patients were like the autumn leaves, clinging desperately to life, but only for a few hours or days, eventually falling ever so sorrowfully into a long sleep.

October 23, 2014: "These Are Hard Times"

Last night death claimed two more lives. Eleven-month-old Beyan, cared for lovingly and tirelessly by his papa for days on end, finally lost his battle against Ebola. He was buried in the early morning. I did not see Papa George, who must have been heartbroken to lose his child. Beyan's toys and Papa George's belt, discarded in the barrel of trash outside their room, were the only cruel reminders of their gallant but ultimately losing struggle against Ebola.

Forty-two-year-old Morris, whom I never saw alive, died after only two days in the ETU. The burial team came to spray him to prepare him for burial but did not first ask his roommate, Sekou, to leave the room. Sekou was sitting bolt upright across from the dead man, staring at him intently. When I chanced upon them, I quickly ushered Sekou into an empty room. I shuddered to think what was going through his mind— first witnessing a death and then seeing what was done to the dead person. I hoped this did not dull his will to fight his illness. He seemed to have read my mind: as

I sat him down on the empty bed, he told me he was strong and he was going to beat Ebola.

Morris's thin body was too long for the bed. He was stretched diagonally across it, facedown, half covered with a sheet and a wrapper, one leg sticking off the railing. The burial team sprayed him liberally, then turned him over and put his arms across his chest. His eyes were closed and he looked peacefully asleep. He was again sprayed thoroughly. A body bag was placed on the stretcher on the floor, the bag was sprayed inside and out, then a second body bag was placed inside the first and sprayed. Morris was covered with the sheet and wrapper except for his face—in case his family wanted to view him before the burial. His body was double bagged and zipped securely. The burial team lifted the stretcher and moved on: a sprayer followed, spraying the path behind them.

Now it was the turn of the psychosocial team to get in touch with his family. One of them said in a broken voice, "These are hard times." She lowered her eyes and suppressed her tears.

The psychosocial team at Bong was a vitally important part of the day-to-day running of the ETU. They dealt with patient mental health, the acceptance of patients back into their villages, placement of orphans, discharges, informing relatives of deaths, and burials. One of the nurses who served on the team also acted as a pastoral counselor and religious leader of devotions and funeral ceremonies.

An ambulance brought in Mason yesterday afternoon. Thin and emaciated, he was admitted into the suspected ward after being triaged in the field by one of the doctors. He coughed up blood this morning and evidently had been doing so for two months, even though he'd had no exposure to Ebola patients. His body was so wasted you could see his rib cage covered only thinly by taut skin. I peered into his mouth but did not see thrush, which would have indicated possible HIV coinfection. He coughed up so much blood I asked to have him moved to the far corner near the exit, away from the rest of the patients. Tuberculosis seemed more likely than Ebola; he probably should not have been admitted to the ETU. Indeed, he later tested negative for Ebola and was transferred to Phebe Hospital down the road—a teaching hospital for Cuttington University that lost five of their six nurses to Ebola in June and had to shut down for a spell. Understandably, the staff was now cautious about accepting any patient potentially infected with Ebola and was running only at a third of capacity.

In the triage area we interviewed Payne, a teacher who lived in a very rural area and had become concerned when his chronic stomach pain seemed worse. He decided it would be prudent to be seen at an ETU, so he traveled across the river and finally reached an ambulance service that brought him to the ETU. The trip had taken him about seven hours. After some questioning we determined that he did not fulfill the criteria for Ebola. Because he had stepped inside the hot zone, however, he had to be decontaminated before discharge. He showered, his clothes were bleached and bagged, his cell phone soaked in high-strength chlorine for thirty minutes (all but assuring that it would be damaged); then he was given a set of new clothes and shoes and transported into the low-risk area, where his blood would be drawn and sent to the lab. Before he could return to his village, Mr. Payne would have to wait for a day for the test result and a paper certifying he had tested negative.

At the end of the day, we had great news to deliver to Alice and Femata: some twenty days after being first stricken, their Ebola tests were finally negative! Both of them danced with joy.

Throughout the day American Apache helicopters and Ospreys took turns droning over the ETU. The United States had deployed thousands of soldiers to Liberia to build ETUs: the sound of these aircraft was a sign that the world had not forgotten this part of West Africa.

October 24, 2014: The Grim Routine

Two more patients died overnight. One was Annie, fifty-two, who roomed with Nuwah, her twenty-two-year-old daughter. She was as tall and slender as her daughter was broad and stocky. Besides vomiting, Annie's only manifestation of disease had been weakness. After her body was removed from the room, Nuwah looked lost and forlorn, pacing the hall and the gravel walk. She could only look on and grieve from a distance through the fence as the burial team carted her mother away to be buried in the cemetery.

The other dead patient was Kumba, a forty-year-old traditional midwife, robust and heavy but significantly weakened by her infection and bloody diarrhea. This woman, who coached and cheered on village women to deliver their babies, never regained enough strength to leave her bed.

For those of us in the swing shift, the time for rounds was two o'clock, the peak of the afternoon's heat. The donning area was steamy. Sweat trickled down my face and body as I was putting on my protective clothing. There was no fan in the donning room, and I was feeling drained and defeated even before I started. Looking around, I could envision my colleagues as nuns in their habits or as Muslim women in their black hijabs—except that the PPE, our space-age suit, was white. The nurse's aides had already gone in to deliver a late lunch. At this time of the day, rounds had to be expeditious so as not to prolong our time in the wards. We plodded through each, quickly assessing the need of the patients for fluids.

Winner's high fever had continued for the last five days. She had been sitting up in a chair the day before, alert and being fed by her uncle while holding a Barbie doll in her left hand. But now, during rounds in the hot afternoon, she had puffy eyes and was found sprawling in her bed, bleeding from her IV site, barely awake. Bendu, who was sure to recover, had been helping her uncle to care for her. She seemed taken by the girl.

Young Solomon was finally recovering. He had begun to show an interest in the world, playing with a truck given to him by the psychosocial worker. There was now a glint of life in his eyes, and he was strong enough to get out of bed and venture outside in the late afternoon, joining a number of the recovering, albeit worn-out, patients who lounged around on the graveled grounds. However, his cousin, six-year-old Christine, was quiet and reserved, refusing her favorite—mashed cassava or *fufu*—as well as fluids altogether. When I asked her whether she missed her mommy, she nodded. I hoped that Communications could devise a way to show her a picture of her mother on an iPad.

Ozona was weakened by bloody diarrhea. Blood oozing from any orifice of the body did not bode well. He teetered on the edge of life and death and began to look like the other patients who had succumbed.

Nancy, tall and erect, carried herself with grace despite her struggle with profuse diarrhea. She had been given tons of IV fluid, but we had no idea what sort of electrolyte imbalances she might be harboring. At first she was able to bring herself to the latrine, but as she weakened she had to resort to using the bedside bucket. Much of the day she looked well, was dressed in a clean smock with her hair neatly done in cornrows, and talked to some of the younger patients in a lively manner; then abruptly she plunged into a troubled well of nausea, vomiting, and desperate weakness. It was impossible to tell which way she was heading.

Today the sweat was literally dripping through my N-95 respirator, tracing its way along my body inside my gown and pooling in the bottoms of my boots. We were hoping to make rounds short, but IV bags and medications all had to be emptied and disconnected before we could leave. We were in the bubble of humidity for close to an hour and forty minutes. When I finally pulled off my layers, I found my scrub pants as soaked as if someone had doused them with water. Thirstily I downed a liter of water.

Before the night shift, the hygienic crew went into the ETU wards to pick up trash for burning. The acrid smell filled the evening air. Meanwhile the ambulance brought in three more patients; one was a woman in her first trimester of pregnancy who had vaginal bleeding. Another was the mother of Morris, the man whose body had been too tall to fit onto his deathbed.

As the crows flew from the morgue toward the leper colony, many of our patients settled into their chairs on the graveled area outside the wards. There was excitement in the air as someone rigged up a sheet to use as a screen while the person from Communications fiddled with some equipment. Suddenly there was a loud noise—music erupted from the boom box and images flickered across the screen. *The Lion King* was the first movie to be shown in the ETU, to be followed by many more. For a brief moment our patients could try on a *hakuna matata* mood to help them in their grim struggle.

October 25, 2014: Children in the ETU and Quarantine News From Home

Children afflicted with Ebola were probably some of the loneliest people in this world. When their test results came back positive, they were wrenched away from their loved ones and led to a blue tarpaulin room, hot and humid during the day, surrounded by strangers, many of them adults groaning with pain. From day one in the ward, they were visited by strangers clad in full PPE, eyes peering through goggles; they heard muffled voices through masks urging them to drink and eat or asking them how they were doing. In between rounds, there was no one to call for help except other sick patients. For days they saw no familiar faces, had no one to comfort them or hold their hands, though a few lucky ones had relatives who had recovered from Ebola and were immune from reinfection, enabling them to reenter the ETU to care for them. Many, feeling rejected, abandoned, and confused, became apathetic and lost much of their ability to fight the infection.

However frightful the first sight of a moon-suited person must be, I had yet to see a child in Africa shrink away from one of us. In America, I imagined, many children would scream their heads off.

The lights in the ETU wards were always on, and it was hard to discern night from day. Because of the heat, the tarp window shades were often down. Daylight only streamed in at the far ends of the long, straight passageway connecting the patients' rooms. If a patient were too weak to move around and sit outside, the hours spent in the ward blended together into one long, endless day, under the harsh glare of the unnatural light.

Seven-year-old Winner was very ill. The swelling around her eyes was less pronounced since we slowed down the delivery of IV fluid, but she was barely awake and cried piteously when moved. The three other children were now rooming together: Joe, Solomon, and Christine. Christine was the sickest. She refused to eat or drink and seemed to be giving up her fight. I wondered if it would make a difference if her mother were still alive. Eleven-year-old Joe, a newcomer to the ward, was moved from a room he shared with young Zonnah, to spare him from witnessing Zonnah as he took a turn for the worse, vomiting up blood.

At one point, I went outside and found Joe in good spirits. Somehow a hint of my smile must have penetrated my Darth Vader suit, for as soon as I approached him, he gave me a timid smile in return. He was sitting on a chair outside reading a book. When I examined the book more closely, I saw that he was reading it upside down. When I righted it and asked him whether he could read, he simply continued to flash his mysterious shadow of a smile.

It seemed to take about a week for patients who were going to recover to feel well enough to get out of bed. It was a good sign when they started moving their chairs to sit outside.

Reports about Ebola and the feared Islamic State of Iraq and Syria (ISIS) dominated the news online. Another volunteer from the US had been infected, heightening the anxiety levels back home. The groundless fear this had generated in the US had resulted in the quarantine of a perfectly healthy returning volunteer for Doctors Without Borders in a tent behind a New Jersey hospital. I was not surprised. Even before I left Boston, I had imagined that as more Ebola cases were detected in the US, public hysteria would mount and any returning volunteer from Liberia would be at risk for quarantine. The politicians were jockeying with each other by promoting hastily rigged plans to gain public favor in advance of the upcoming election.

The days here at Bong were hellishly hot. Today was particularly steamy, and my goggles not only pinched my nose, making it difficult to breathe, but they crunched against my hair clip, giving me a pressure headache. I was not allowed to touch any part of my face with my contaminated gloves. Drops of sweat dripped through my mask and down my sleeves. My mind turned a little hazy. We still had to place an IV line and administer a bag of Ringer's lactate to Sekou.

It was particularly important to me that Sekou received his fluids. He had watched Morris's body being sprayed by the burial team. Each time I saw him, he flexed his biceps and reiterated that he was strong. Looking at his reedy stature, however, one wondered how he would withstand a sudden gusty wind.

I tried to push myself through it all mentally. It was as though I was running a marathon on a particularly hot day. At mile seventeen, I always began to hit the wall but somehow pushed myself to mile twenty. At that point I asked myself why I was running at all. But in the home stretch, the emotional boost of having run 26.2 miles was indescribably uplifting.

And so I pressed on until the last drip in the IV bag was squeezed into the last patient and my partner and I went to the doffing area. When I finally emerged from the inferno of two hours inside the wards, I was blinded by the bright sun. A few puffy white clouds dotted the turquoise sky. My oversize scrubs and clunky boots were completely soaked. At least during a marathon you could stop to hydrate yourself; now I gulped down a good liter of water.

Right outside the donning area was a whiteboard where monitors recorded the names, entry, and exit times of the persons entering the wards. However, there were no clocks inside the wards for us to monitor how long we had been there. In most instances, there were so many sick patients we did not feel we could leave without their receiving their specified amount of fluids. However, it could also be foolhardy to overstretch our time in the ward.

Over the orange net fence, I could still see Joe

looking intently at his book while his former roommate, Zonnah, was probably cooped up in his room, struggling alone with his nausea and vomiting.

October 27, 2014: "O Death, Where Is Thy Sting? O Grave, Where Is Thy Victory?"

One could almost predict which patients would not do well: the ones bleeding from the mouth, and nose.

Winner and Zonnah had finally succumbed. They were both quietly buried this morning. I wanted to attend Winner's burial, but my morning rounds took precedence. In the afternoon, when there was a lull, I took the lonely walk on the dirt path around the ETU, through the shady forest to the cemetery. I found Winner's wooden tomb marker. She was buried under a canopy of overhanging bushes, a sheltered area. I prayed softly that she might rest in peace. All was quiet; the grave diggers were gone for the day.

There were now forty-two patients buried in the cemetery, and the ten grave diggers had already dug more graves: mounds of fresh brown earth were piled up beside a row of neatly measured holes. Many grave diggers were shunned by their fellow villagers, who feared that they might bring Ebola to their villages, and so they lived apart from their families while they were doing this necessary but deeply sad job.

I had been switched back to the seven A.M. to seven P.M. shift. At sign-out this morning, for a change, there was no one in the suspected ward. Two days ago we admitted to the ETU the last seven members of a family of twenty; each had had contact with a sick relative who later died. Five of the seven patients tested positive for Ebola. Two of them, Patience and Dorcas, had some of the highest numbers of virus in their blood seen by the lab.

Rounds in the ward now started for me at eight forty-five in the morning, a far more pleasant time than the afternoon as far as heat and humidity went. Unfortunately the fogging of my goggles still dogged me, making it difficult to read patient charts, fill out forms, or order medication and IV fluids.

Duo, a tall, muscular Liberian physician's assistant, went on rounds with me. When we stepped into Nancy's room, we found her nearly naked, sitting on the bucket commode and hunched over her bed, face lowered in her arms, her cornrowed hair draped over her neck and upper arms. Weakness had decimated her once graceful body, but somehow she had succeeded in hauling herself to the commode during the night. When we touched her, however, we realized that she was already in rigor mortis. She had expired while sitting on the bucket. I stared in disbelief. The day before, she had managed to walk to the gravel area up front and socialize with a few of the healthier men. We picked her up onto the bed and attempted to straighten her body. Duo went in search of the burial team.

In the wards, there was no bell to ring for a nurse. Unless a fellow patient yelled across the orange net fence for help, a needy patient just had to wait for the next shift to come around. Even in such an emergency, it would take twenty minutes for help to arrive because of the meticulous donning process. Nancy could not have rung for us in her final moments, and we could not have come quickly enough even if she had.

Patience was quite confused this morning, wandering around listlessly and getting into bed with other patients in a search for comfort and companionship. We led her to her own bed and continued with our rounds. Some time later, while examining another patient, we noticed in that patient's bed an unexpected lump. Pulling off the cover, we discovered Patience curled up in a fetal position. This time we let her stay in the room to which she seemed to gravitate most often. The team that did rounds later in the afternoon reported that Patience had experienced a generalized seizure. We seemed helpless to stop her downward spiral.

In the evening, just when our shift was ending, the ambulance brought in five patients. I was exhausted and wished to go back to my dorm, but we felt compelled to stick around and help out. One of the patients was an eighteen-year-old woman with postpartum hemorrhage who, sadly, died on arrival. Another fallen autumn leaf.

October 28, 2014: The Burial

Patience, the wandering patient with one of the highest Ebola titers, died overnight and was laid to rest this afternoon after only one day in the ward. In the morning I raised the blanket that covered her face. Her eyes were closed and puffy, and the rivers of blood that had oozed from her nose had coagulated. It did not look like an easy death. She was only nineteen.

The morgue was a long blue tarpaulin building with no windows, only flaps for entrance and exit. It sat quietly at the western end of the ETU next to the confirmed ward. I walked down a sloping gravel path, then up a cement walkway. Gingerly I picked up one end of the entrance flap and peeked inside. A single shrouded body rested on a stretcher on the concrete floor, ready for burial. Despite the heat, it felt cold, desolate, and lonely. Like a voyeur, suddenly I was overcome with the strong, tumultuous emotion that I had intruded into the intensely painful and private final journey of this faceless person. I let the flap drop gently, turned, and walked down the walkway.

Nancy's sister came to see her laid to rest. The burial team carried her body on a stretcher from the morgue to the gravesite along a wide dirt path the sprayer spraying the path just taken; meanwhile we took a meandering route through the shaded forest, with Nancy's sister carrying the grave stake bearing Nancy's name. At the cemetery, she was asked if she wanted to view Nancy's body. She gave an imperceptible nod. The double body bags were unzipped to reveal Nancy's face for a brief moment; the open mouth showed off her white teeth. Her eyes were closed. It was not the most flattering final hour for Nancy, who had taken meticulous care of her appearance even when gravely ill. Her sister quickly averted her eyes. No loud wailing or shedding of tears, just quiet grieving. Ebola had put an end to touching and hugging during last goodbyes.

A Liberian nurse who was also a psychosocial worker said a short prayer. Then he and Nancy's sister turned slowly away and retraced their heavy steps toward the ETU.

Methodically, the burial team tied both ends of the body bag with bright-red heavy plastic cords and lowered it into the grave, throwing the cords in after it. The sprayer drenched the stretcher with chlorine, and other members of the burial team took it away. The grave diggers waited while I peered into the grave at the shrouded body, then efficiently but reverently began filling in the dirt.

In the late morning, the US ambassador to the United Nations, Samantha Power, came for a brief but polite visit. I imagined she was helicoptered into Suakoko to avoid having to take the long, bumpy, difficult roads. After all the trouble taken to come here, she turned down an invitation to step into the low-risk zone and instead she received reports from the ETU leaders in the green zone. The visit was really just a whirlwind photo-op tour. The whole entourage gave the ETU a wide berth, viewing it from about a hundred meters away; none would even touch the fence that separated the green zone from the low-risk area. At first one enthusiastic reporter led the way, stepping in her boots into a footbath in preparation for entering the low-risk area. She was called back by the other visitors. After that, all reporters conducted their interviews of health care workers from across the fence.

On this same day, a group of United States Public Health Service (USPHS) doctors and nurses came to the Bong ETU for so-called hot training. So I made rounds twice, first for over two hours in the morning, and then for an hour in the midafternoon with the USPHS doctors. Several of these doctors and nurses had attended the CDC course with me. It was great seeing familiar faces—we had a sort of a camaraderie having gone through the simulated course in the US. Here in Bong, I felt a little more seasoned, since I had been through the real deal now and would be their trainer.

After they completed their training, they would go work at the hospital being built near the Monrovia airport by the US Army to care for both expat and national health care workers who had contracted Ebola. The whole ETU would be air-conditioned, and their PPE was supposed to have a built-in ventilation system to keep them cool!

I took the new doctors into the ETU and showed them around. Christine, the six-year-old, was sitting up and eating a piece of chicken. She seemed to have turned a corner. Duo, the physician's assistant, made a beeline to her room, picked her up and walked her along the corridor, holding her little hand. Duo had children of a similar age and had taken a liking to Christine. In the afternoon she sat outside in the shade with the two recovering boys; her cousin Solomon and Joe, the eleven-year-old who had rebounded quickly from his illness. Two other young boys, Alfred and Munyah, had come in last night, quite ill with confirmed Ebola.

Patients who became symptomless for three consecutive days were tested daily to see whether they were truly Ebola-free. For some time now Bendu, the young woman who arrived here three weeks ago and helped care for poor Winner once she began to get better, had been hoping to have good news. Late in the afternoon when the test results came in, Bendu's test finally turned up negative! She screamed with elation that she was free from Ebola and danced wildly.

The ambulance, however, continued to bring in more patients. Four came at the end of the day, one of whom had died en route. Watching the cleaners and sprayers do the dangerous job of moving a body already in rigor mortis through the tight space allowed by the makeshift ambulance, I could not help but think about the risks they were exposed to daily. The nationals who chose this line of work had a great deal of courage. Like the grave diggers, some were pariahs in their own villages, shunned by those closest to them, and forced to live away from home. . . .

October 29, 2014: Lives Touched by Ebola

It seemed that almost everyone here who was affected by Ebola had lost someone in the family to the disease. A staff member relayed to us that in a small village of sixty-eight inhabitants in the Suakoko district, a quarter had fallen ill with the virus, with only seven of those surviving.

Watta's son, eleven-year-old Alfred, had Ebola while she was negative; he had been exposed to his brother, now dead. He and his mother were admitted together yesterday evening and went into separate wards: he was now in the confirmed ward and she in suspected. Watta called to him in the morning over the fence dividing the two wards to check on him before she showered and was discharged. She left reluctantly, lingering and turning around in the hope of catching a glimpse of her son. Alfred was rooming with fourteen-year-old George, who was similarly ill, but with a mother who had already perished from the illness.

The sickest person was fourteen-year-old Munyah, who had been bleeding from his gums and refusing to take his medications, perhaps because of the pain in his mouth. His grandmother had died from Ebola.

Esther's death a week ago took us all by surprise. She

had been looking perfectly well, but it turned out to be just the calm before a storm. And yesterday morning, Esther's daughter Satta was brought in during the afternoon and confirmed as infected. She was now in the confirmed ward, alone without her mother or relatives or familiar faces. Instinctively she shrank from all human contact and withdrew into herself, crying when anyone paid her the slightest attention. Nuwah, the young woman whose mother had died a few days ago, was improving while she waited for her Ebola test to turn negative, and she had been recruited to take care of Satta.

Andrew, whose wife had died from Ebola, was doing well, and said emphatically that he was going to win his fight against Ebola.

Young Solomon and Joe would recover; they were now quite rambunctious and tagging along behind us in our rounds. Joe followed me into the storeroom to ask for another pair of slippers; one of his had a hole in the sole. All I could find him was a pair of flip-flops that were a tad small.

Christine had lost both mother and brother, but today her father came to visit, bringing her oranges and soft drinks. The visitor hut was partitioned down in the middle by a wall and wire mesh; visitors were physically separated from patients by about two meters. The nurse carrying Christine into the hut was hugging her, so Christine did not know her father was there until he spoke and she recognized his voice. She turned around, stretched out her arms toward her dad, and began to cry, "I want to go home." Her father told her she had to eat and drink well and get stronger before she could come home. She nodded. A few days ago, when she asked for her favorite foods—cassava leaves and pawpaw—we had the feeling that she might be getting better. But sitting there at the table today, she looked fragile and frail.

Outside, the sun baked the red earth and the gravel covering the grounds of the ETU multiplied the heat of its rays. We made rounds this morning with the USPHS folks. With so many trainees, monitors had to slow down the doffing process to avoid deadly mishaps. The resulting jam meant that some of us were stuck in the wards longer than planned.

In the late afternoon, there were so many visitors and trainees in the office I found no room for decompression. I went to find some quiet on the back steps of the changing rooms. I was awakened suddenly from my reverie when someone yelled, "A snake, a snake, kill it!"

A greenish-brown three-foot snake had slithered out of the stockroom next door. It slipped under the fence, and in a flash its undulating body crossed the red dirt road in desperation as a piece of wood hurled over the fence narrowly missed it. One of a pair of grave diggers walking up the path at the end of their workday ran toward the snake in hot pursuit.

"Don't kill it, let it go!" I yelled at him.

But the snake was faster than the man, and it disappeared into the thicket of cassava bushes before he could reach it.

"Why don't you want me to kill it?" the man asked. "I can eat it."

I did not want to say that I had already seen too much death, so instead I said I was afraid the snake might turn around and bite him.

"Oh," he said. "Thanks for coming here to fight Ebola with our families."

October 30, 2014: Confusion in the ETU

Early this morning Alfred, who was now the sickest kid in the ward, managed to wander barefoot into the low-risk zone, where we health care workers normally congregated. He was disoriented and confused, walking on the graveled compound like a zombie, arms stiff at his sides, body bent forward. We decided he must have sneaked through the doffing zone undetected. Everyone stayed as far away from him as possible while a quick-thinking nurse went quickly to put on PPE. At first he stood rooted in place, but then he was so weak that he slowly slumped to the ground, first on all fours, and then, some minutes later, he could not sustain this stance any longer and he slid down in slow motion onto the gravel and curled up onto his side. The scene was heartbreaking: he had given up his wandering quest. The nurse in full PPE picked him up reassuringly and walked him back in painful steps through the doffing area and back to the confirmed ward. All the while the psychosocial nurse was leading a devotional in front of a group of confirmed patients across the orange net fence, unperturbed—as though this kind of thing happened every day. Most of the patients sat slumped in their white plastic chairs, looking tired and worn-out.

We tried to recruit the two older men in the ward to stop Alfred from wandering, but they were not interested. Perhaps they were too exhausted themselves to care for him. Solomon and Joe, who had ample free time on their hands, were also asked, but they were more interested in fooling around than in babysitting. Actually, they had started going to the storage room to get surgical gloves and chlorine spray bottles so they could clean up after George when he threw up his ORS. The boys were slated to have their Ebola tests repeated, since they had been without symptoms for the requisite three consecutive days.

Bendu, who went home two days ago with her wide, happy smile, came back today to be in the confirmed ward with Satta, Esther's three-year-old girl, who had been refusing to eat or drink. She would be her caretaker.

For many days now, I had been unable to end on a happy note. The ambulance brought in two people. We took care of the sicker of the two, Abraham, a thirty-year-old man who'd had symptoms for eight days. He was so confused that it took three of us to literally pull him down from the vinyl mattress in the ambulance and into the stretcher. He was very dehydrated: thin and gaunt, with haunted, sunken eyes and hollow cheeks. It took multiple attempts before we could fix him up with a very small IV line and squeeze into him five hundred milliliters of fluid. We then rigged up a makeshift restraint with strips of bedsheet across the bed rails to prevent him from heaving himself off and wandering around the ward. Given the wasting evident from his sunken temples, he must have had a chronic illness before catching Ebola. The night crew would continue hydrating him, but I feared he would not make it through the night.

A twelve-hour shift often stretched into a fourteen-hour one, but I was particularly exhausted by the end of today, probably on account of the recent influx of sicker patients. We left so late that the patients had already finished watching their nightly movie. I was glad to leave the noisy ETU for my dormitory room, where the silence of the night was only broken by the crickets' enthusiastic, melodic creaking.

October 31, 2014: The Crow on the Roof and the Dancing Boys

Abraham, the man we helped into the suspected ward yesterday, died last night. We never got a blood sample to test for Ebola, since we were too busy trying to get an IV line into him. During his illness he had first been living with a bunch of people and then was briefly in a health care facility, which meant that if he did in fact have Ebola, he might have exposed many people. So we spent some time in the morning trying to obtain specimens from his corpse. First I put a needle into his heart, but instead of blood I only managed to withdraw a copious amount of straw-colored fluid, probably from the heart's lining. The nurse and I thought this should be adequate, but the medical director called over from the donning room said he wanted us to take a blood sample from the femoral vein. This I did successfully, with thick, dark blood coming out promptly when I stuck in a long needle and syringe. The nurse admonished me to be very careful as I transferred the blood into a test tube.

However, this was again not enough. The medical director wanted us to get a piece of skin as well. We finally succeeded, but only after a great deal of effort,

since we had no appropriate biopsy kit—not even twee-
zers or scalpel. Instead we had to use a pair of blunt,
rusty scissors to saw a snippet of skin from the neck.
As we did this, we reminded ourselves that a dead
body contained high levels of Ebola virus. Duo, my
physician's assistant, sprayed 0.5 percent chlorine all
over Abraham's body, the specimen containers, and
our gloves before and after the ordeal. The nurse then
hurried us off to the confirmed ward lest the medical
director ask for more specimens, exclaiming, "Let's go
before he demands a piece of the heart or liver!"

That afternoon Abraham's blood test came back
positive, while the other specimens gave indeterminate
results. Now there would have to be a lot of contact
tracing, and I was glad we had succeeded in drawing
some blood.

Peewe and her mother came in to the suspected
ward yesterday. Her mother insisted on keeping her
company, even though she herself was not ill. In the
afternoon, when I told Peewe from the donning room
that her first Ebola test had come back negative, mother
and daughter praised the Lord with "Hallelujah!" and
began to sing. A second test came back later today,
also negative—seventy-two hours after her symptoms
started! On hearing this, both women started to dance.
Soon they would go through the ritual of discarding
all their belongings and cleansing in the shower room
before returning home.

At noon a crow stood on the roof of the morgue,
cawing ominously. When I was young and growing up
in Malaysia, my mother used to say that when a crow
settled on the roof of a house, a death would soon

follow. Last night we had one death, and in the morning we had another two. One was Dorcas, Patience's sister, who like her sibling had had one of the highest Ebola titers in the Bong ETU. She had gallantly kept herself clean daily despite profuse diarrhea and weakness. I rooted hard for her two days ago when she was cleaned by an aide and helped into a sundress, after which she dragged her small frame outside in front of the ward to join some other patients. Sadly, she had lost her battle.

The other fatality this morning was Alfred, the confused, wandering boy. His mother, Watta, who had been calling daily to inquire after her son, came to the burial this afternoon. She had now lost two sons. She did not cry as she gazed into his face when the burial team unzipped the body bag. He looked more at peace than when I saw him this morning after the end of his life's short journey. The burial team picked up the stretcher and followed the wide path to the cemetery, while we took the narrow, shady one. Watta's head was wrapped in a headdress and she wore her wrapper around her waist like a sarong. She was carrying the wooden stake bearing Alfred's name.

The psychosocial nurse led the burial team and grave diggers as they sang,

When we all get to heaven,
What a day of rejoicing that will be!
When we all see Jesus,
We'll sing and shout the victory!

The nurse said a prayer, and Alfred's mother echoed with a resounding "Amen." Alfred's little body was slowly lowered into the grave. Then Watta turned and departed quietly.

Back in the confirmed ward, the day was coming to a close. Despite the tragedies, Solomon and Joe had been performing a dance they created for anyone who would watch. They turned up the radio to full blast and gyrated here and there on the gravel. They gave a particularly exuberant performance this afternoon when the navy lab man, who had children the same age, came personally to tell them their Ebola tests were negative. In unison they shouted, "I am free of Ebola!"

I would miss them after they left in the morning. Tonight would be their last movie night at the ETU.

The melancholy Sekou was standing nearby when the boys received their good news. Everyone was so engrossed in the boys' reaction they at first forgot to tell him that he too had tested negative. He smiled quietly. He had seen his roommate die and be sprayed, but he had told me that he would beat Ebola, and he had!

November 2, 2014: Cuttington University and Gbarnga

A rare day off.

I had been working for over a week according to a nonexistent schedule, regularly switching to different shifts. The coordinator usually told me my shift only the evening before. One Saturday I asked her whether we were ever given a day off, to which she gave the derisive response, "No one takes a day off during an emergency!" And added, "I haven't had a day off!"

I might well have complained to her that though I arrived two weeks ago, I had yet to be given a proper orientation. It took me over a week to find out about food, transportation, and the specifics of medical rounds. More importantly, all incoming health care workers were supposed to be given a thermometer and instructed to take their temperature twice a day while working in the ETU. That was never conveyed to me. I had to go to the pharmacy myself to ask for a thermometer after Peris explained the protocol.

After some time, the coordinator reluctantly allowed that I could take the next day off, which was a Sunday. "Why did you want the day off, anyway?" she remarked. "Markets are all closed on Sunday!"

I held my tongue. Who said I was going to the market? Besides, it was not true that no one took time off. Almost all those working here, with a few exceptions—including me—were on the IMC payroll. Every one of them followed a schedule that factored in off days. The coordinator might very well have worked every day, but if so that was her choice, not obligation.

The grounds of Cuttington University had been empty of students since the Ebola outbreak. One of the oldest private colleges in sub-Saharan Africa, the campus was crisscrossed with red dirt roads carved with deep gullies by the rain; overgrown grass threatened to choke out everything else. Small buildings were scattered across the rolling green landscape; dormitories or guesthouses seemed to predominate rather than classroom buildings. There were even a few ruins and a defunct museum, courtesy of civil wars that had ravaged the campus.

Sleep after a night shift was nearly impossible because of the yelling and loud noises from the hall that streamed through the vents during the day. Luckily a few days earlier the departure of some volunteers freed up some housing space, and Peris and I were moved into one of the guesthouses. We now shared a bedroom but had a big living room, dining room, kitchen, and a bathroom, and all in all it was much more comfortable than the dormitory. Here we also had a working refrigerator, and the kitchen was stocked with sugar, cocoa, tea, and coffee, though we lacked the least semblance of any cooking utensils. As in the dormitory, there was no running water in the bathroom, where a big barrel of water took up much of the space. There *was* running water in the kitchen, however. I requested that they move the fan in my dormitory room to the guesthouse, since the heat was unbearable.

On the morning of my day off, I went to visit the Ebola lab the US Navy set up on campus within ninety-six hours of their landing in Liberia. The building that housed it had been spruced up, with ceilings patched and a fresh coat of paint. The lab equipment requires air-conditioned rooms, so, since the arrival of the navy, the campus had enjoyed electricity around the clock instead of from seven till ten at night. They could process ninety-plus tests within four hours in the building's biosafety lab, the hot zone where PPE must be worn, and that was far from full capacity. While the rest of the Cuttington campus did not inspire much confidence, this lab was a state-of-the-art operation.

Next door to the Navy lab was the storage room for International Medical Corps. About fifty Liberians were gathered in front of the building on this Sunday morning. I later learned that they congregated there every day looking for a job with IMC.

I walked to the campus church—Epiphany Chapel—for their ten o'clock Sunday service. The sun was beginning to bestow its heat, and I had on my broad-brimmed sun hat. I sneaked in from the back but did not escape the attention of a lady usher clad in a slinky traditional dress. She handed me a Book of Common Prayer whose cover was sticky and its edges brown and curled. I took my hat off and spread out on one of the benches in the last few rows. At capacity the chapel would hold a couple hundred; today there were about fifty, mostly clustered in two or three sections of the church.

Strain as I might to understand the priest, I had a hard time piecing together his sermon as he droned on in his Liberian English. Small-paned glass windows lined the sides of the chapel from floor to the ceiling. Only a few had been flipped open, and no breeze came in to cool us. The ceiling fans were not turned on or perhaps did not work. Sweat trickled down the inside of my dress. When we exchanged the peace of God, we just waved, without touching or hugging. There was no communion, and none would be served until sometime in December.

Suddenly a few congregants stood up, and it dawned on me that the priest wanted to single out all the newcomers. Having missed my cue, I decided to stay seated. I was the only foreigner, and I already felt too conspicuous. We sang without keyboard accompaniment, and the result was on the funereal side. During the offering Samuel, one of the helpers, who also worked at the ETU, beat out a rousing rhythm with the drum that was pinned between his legs. It was the only lively thing during the long service, and it made me appreciate all the more the music I enjoyed at my church at home.

At the end of the service, the priest stood at the door greeting us. Again, no hand shaking.

In the afternoon one of the logisticians had to run some errands in the nearest town, Gbarnga, and he took me along. We stopped at a street-side market where women were selling produce—oranges, potatoes, and one big, tempting pineapple, which the logistician bought. Oranges were in season. Colorful umbrellas fluttered in the distance. We walked to an Ebola check-point to have our temperatures scanned. Outside, next to a colorful banner that loudly declared, "Ebola is real," a mother and daughter dutifully washed their hands in chlorinated water.

Gbarnga was like many African towns, with unpaved roads and sprawling markets that sold all sorts of stuff. Even on Sunday the market seemed busy (despite what the coordinator had told me!). The logistician, who was Dutch, walked through the market attentively. In the not-too-distant past, he had been harassed there on account of being a foreigner. We ended up in the Total Gas station, where the proprietor had a supply of Western food: digestive biscuits, chocolates, liquor, juices. Nothing I was interested in . . . until I found a *minimum buah laici*: lychee fruit juice, a product of Bangladesh but, curiously enough, with labeling all in Malay, the language of my country of birth. For nostalgic reasons, I purchased it with my small stash of Liberian dollars.

WHEN THE US NAVY came to Cuttington University to set up the laboratory for the testing of Ebola, the equipment had to be in an air-conditioned room.

The bonus for us was we were able to have electricity and Wi-Fi 24-7. About five years ago, I had started to blog whenever I volunteered as a medical doctor as a way of keeping a journal and providing an avenue for my family and friends to follow my experiences without my having to write to them individually. It also offered them some peace of mind as to my safety in some dicey situations in foreign lands.

I was finally able to coordinate a time to Skype with my family when my husband and my three children could all be present on a Sunday. After my jaunt in Gbarnga on my day off, I collected my boxed dinner from the kitchen on campus, went back to the guesthouse, and washed up to make myself comfortable to spend an hour catching up with my family. It was a relief for them to see me still whole and intact on Skype and for me to talk to them after such an intense two weeks in the ETU. They were particularly worried about the two instances of my potential exposure to contaminated fluids. I realized that in relating them in my blog, instead of allaying anxiety, I had caused them unnecessary fear and turmoil.

After a long day off, it was wonderful to reconnect with loved ones again. I felt truly rejuvenated and energized to return to fight another day for my patients, who were battling a far graver war than mine would ever be.

November 3, 2014: "Like Moths to a Flame"

After my day off, I received a text message from the coordinator that I was being reverted to night duty. What with deaths and discharges, the census in the wards was down. Rumors were that Ebola cases in Liberia had decreased, but when this had happened a few months ago, it was soon followed by a resurgence. It took only one case of unsuspected infection to cause an outbreak.

Alfred's mother, Watta, returned to be readmitted. The cause was nothing more than a headache without fever, but she was understandably worried. Luckily she ran no danger of close contact with infected people, as she was the only patient in the eerily empty suspected ward. She tossed and turned all night, unable to sleep under the twenty-four-hour lights.

There were fourteen patients in the confirmed ward. One of them—Sial, whose brother Morris died a week ago—was critically ill. The physician's assistant, Vasco, and I went into the ward in the early evening and before dawn to check on her. She had been vigorously hydrated. At dawn we found her in a pool of loose stool, completely soaked and shivering with a high fever. We could rouse her only by applying pressure on her sternum, which made her groan faintly. Vasco and I were there without nursing staff. Cleaning and changing Sial and her bedsheets between the two of us was inordinately difficult. Her elderly mother, Tewah, herself a patient lying across the hall, heard us fussing over her daughter and wanted to know how Sial was doing. Tewah had been struggling with a distended abdomen for several days now and in the early morning hours began to have bloody stool. I feared that, having recently lost a son, she would soon also lose her daughter.

Seeking comfort, three-year-old Satta crawled into bed to sleep with Christine, the six-year-old; naturally enough, she gravitated toward her own age group. A number of our slightly older youngsters had bounced back, including Munyah, though he had experienced some slow bleeding from his mouth and nose.

During the night we pushed together the round plastic tables, swept them clear of tiny insects and Nairobi flies, and tried to catnap on top. I folded a sweater as a pillow and covered myself with a long shawl, but a few mosquitoes tried to take advantage of our netless situation. Loud conversations and the blaring soundtrack of Nollywood movies made me wish I had earplugs; the Disney movies that patients complained were boring had been replaced by more adult fare. By this time of the night, however, it was only the Liberian health workers watching, as all patients had retired to bed. They were mostly talking among themselves, not really paying attention to the screen. I got up and gestured to them that I was going to turn down the volume. The staff spoke Liberian English and I rarely understood a word: the speech pattern seemed to lack intonation or modulation. Often I mistook it for the local dialect.

I lay back down and listened to the sound of the insects. To be in this place must be an entomologist's field day. The varieties of moths were endless—tiny and enormous ones, some nondescript shades of brown the color of dead leaves, and others snow white with speckles of green and black. Early dawn revealed the carcasses of dead moths piled up beneath each energy-saving lightbulb. I immediately thought of our critically ill patients, drawn to death like moths to a flame.

November 4, 2014: The Hiccups
All hell seemed to break loose yesterday when we arrived for the night shift. Consistent scheduling of health care

personnel had been lacking since I got here, and it turned out that one of the night-duty nurses last night had been switched at the last minute, leaving us one person short. Meanwhile, our census soared to twenty-eight patients: six more patients brought in by our own ambulance and another seven sent over by Phebe Hospital. These last consisted of two families: two mothers with young children; one of the babies, named Moses, was moribund.

In the confirmed ward, James, a disoriented forty-nine-year-old who looked much older than his age, had an expression of anxiety, fear, and panic in his eyes. He was also hiccuping regularly. I had noticed that a number of our patients had the hiccups, and I couldn't help but think of it as an ill omen. We pushed a liter of fluid into him.

A Kenyan nurse named Paul and I went on rounds again in the early hours to check on the very sick patients. The baby Moses continued to be unresponsive and began to have bloody diarrhea in his diaper. His mother, Krubo, looked on with despair while two of her children slept soundly on the next bed. We pushed fluid through a small IV line into his leg using a ten-cc syringe. The father had died of Ebola not long before.

Sial, who had been gasping for air near the end of the day shift, finally breathed her last. When her mother, Tewah, had asked after her earlier, the nurse had replied that she was sleeping—a euphemism for the process of dying. I felt that Tewah might really want to be given a last chance to say farewell or even touch her, but the nurse insisted that Tewah was too frail and, being located right across the hall from her daughter, might wail and cause a ruckus. When we last checked in the wee hours of the morning, Tewah was so sound asleep we did not have the heart to wake her. Instead, we decided to have one of the psychosocial nurses break the news to her.

When we passed by James in the last bed in the far corner of the confirmed ward, he was still hiccuping in his restless sleep.

The muezzin chanted and prayed in the early morning while the cock crowed. Dawn broke slowly through the thick fog while Sial "slept" in her bed with one arm across her chest. She would soon join her brother in the cemetery.

November 5, 2014: All the Perfumes of Arabia

Sadly, close, unprotected contact with Ebola begets more Ebola. The mothers and children of the two families that we admitted last night were all positive except for the two-year-old boy, Jackson. He refused the formula that we made for him when we admitted him to the suspected ward and demanded to be breastfed by his mother, Fatu. Our coordinator, who was not a medical doctor, insisted that it was all right to let the baby breastfeed. Most of us disagreed, but she overruled us. Once the test result came back negative, however, he was quickly switched to formula.

But that was not the end of the problem. While we waited for a family member to come take him home, the same coordinator made the decision to move Jackson into the confirmed ward to be with his mother, even though this put him at risk and the more prudent thing to do was to have Bendu care for him in the suspected ward. I could not understand the reasoning behind this decision, which was made in my absence. Fatu's other child, Theresa, had the highest viral titer in the region. Otherwise she looked well, except for having a fever. I feared that this might just be the calm before the storm.

Moses's mother, Krubo, seemed to take little interest in him, and he remained unresponsive. He occupied the center of their shared bed while Krubo had scooted over to the railing as far away from her baby as possible. Once she even gestured at me to take her baby away, though he had yet to die. The other two children— Faith, who was four, and Peter, who was five—were having gastrointestinal symptoms and fever; they were struggling to stay afloat.

James's hiccups had stopped but he continued to be delirious, answering to the name of the late Libyan dictator, Gaddhafi. I did not know why the staff seemed to think that's what he wanted to be called. I always called him James.

Some delirious patients began to pick at their sheets. There was deep fear lurking in the pools of their eyes, and always a desperate, silent plea for help—one for which we had no solutions. We squeezed a shoulder or a hand, but in the end we walked away offering very little. The fear remained. The cry for help persisted.

The distension in Tewah's abdomen pushed against her diaphragm, and she had trouble breathing. Last night she was wheezy and restless. Lying down made her even more breathless. Someone found her a pillow—a rare commodity in the ward—and propped her up. That eased her breathing, but she still needed to get up and pace the floor well into the night. We searched for dexamethasone in the pharmacy, without success. She was given a lot of fluids during the day; her body had been unable to get rid of the excess fluids that accumulated in her belly, making it difficult for her to breathe. We had little to offer her—no oxygen, no steroids, no nebulizer, no water pills. The family came to attend her daughter Sial's burial. The bananas, papaya, and cucumbers they brought sat beside her bed, untouched.

Last night after we doffed we discovered that all the taps for 0.05 percent chlorine and some of the taps for water were bone-dry. Only the taps for the strong solution of 0.5 percent chlorine were running. It turned out that the pumps had been inadvertently turned off after the day crew left. Chlorine was our armor against Ebola, but the strong stuff was too much of a good thing. Nevertheless we had to wash off with it, then quickly rinse off with what water there was. The fumes permeated and stung our sinuses.

When I was back in the guesthouse, I sprayed my boots liberally with my 0.5 percent chlorine spray bottle. Despite the thorough bathing and hair washing, I was sure that all the perfumes of Arabia could not wash away the smell of chlorine that clung to my hands and body.

November 6, 2014: The Untouchables

For some days Naomi and her daughter Josephine seesawed between getting better and looking worse. They shared a room, and every day that we checked on them we wondered whether they would pull through or not. Then one morning when Josephine refused her medicines, Naomi picked up a broomstick and whacked her daughter so severely that the physician had to intervene before she sustained serious injuries. We realized then that Naomi was getting better. Yesterday their long-awaited negative Ebola tests came back, and their departure from the ETU finally took place.

The psychosocial team had set up a memorial wall on the outside of the doctors' office where patients who had become free of Ebola left their palm prints and initials before they departed from the ETU. A small triumph.

Baby Moses finally died. His sister, Faith, now shared her mother's bed, curling up beside her. For her part, Krubo looked quiet and resigned.

Fatu, mother of the toddler Jackson, had started to have bloody diarrhea. Jackson, who was Ebola-free, had finally been separated from his mother to be cared for by Comfort, a patient who recovered from Ebola and was also a nurses' aide. They would stay in a new holding area built outside the fence of the ETU. Fatu's daughter, Theresa, had the "hot body" and "runny stomach" and looked dazed from her illness.

The 0.05 percent chlorine stopped flowing again even earlier during our last night shift. This time the valve to the backup tank had been shut off. Frantic calls were made to the WASH manager, who had to be roused from his dinner. He came and fixed the problem. We had to delay our rounds for an hour until it was turned on again, for fear we would not have enough chlorine for doffing.

James, the patient with the hiccups, had not left his bed since he was admitted. He asked for water in the wee hours of the morning as Duo and I made our second round close to dawn. When I tried to help him sit up, he leaned his full weight against my right arm and it took a behemoth effort on my part, despite his skinny frame. He gulped down a whole five-hundred-milliliter bottle of water very slowly as we gave him another half a liter of Ringer's lactate through his IV. Then we tucked him into bed.

Some of the physicians had completed their stints and departed from Bong, and there were now only two physicians left besides the medical director. The two of us divided the day and night shifts between us.

For some time I had wanted to see how the ambulance crew brought in patients from the villages. Since no physician could be spared the rounds of duty, the only way for me to do that was to go on the day I switched shifts instead of resting up for the next shift. I could not find any way around it, so after my last night shift, I went along with the ambulance crew to Kakata in Margibi County, about two and a half hours away, to pick up some patients.

In the end I was disappointed, as we did not go into any actual villages. It was the Liberian Thanksgiving holiday, and there were only two patients, who were already at the hospital in Kakata, to be picked up. The procedure worked like this: first, suspected Ebola patients or contacts of known Ebola patients in Margibi County were identified by the district council of the Liberian Ministry of Health. If the patients lived too far away, they were brought via ambulance to the isolation units in Kakata's C. H. Rennie Hospital (which had lost more than twenty personnel to Ebola), and our ambulances picked them up from there. The ambulances were actually nothing more than pickup trucks rigged with tarpaulin over some metal frame. We traveled with four of them in case there turned out to be a large group of patients. Each ambulance could carry up to six patients. However, to minimize exposing patients, the crew tried to assign only one to each vehicle unless the patients were from the same family. I was riding in a Land Cruiser with Elvis, the chief of the ambulance crew, tailing the ambulances. He communicated with them with a walkie-talkie.

When we finally got to Kakata, the two male patients were sitting in the hospital ambulance with its back door open for ventilation. The sprayer sprayed one of our ambulances inside and out with chlorine. The men, who were able to walk, climbed down from the first ambulance and up into ours. Since they could move on their own, our crew did not have to put on full PPE to help them. Everyone else made sure to stand some feet from them. Throngs of people lined the street gawking, as though the men had done something wrong. What fear and stigma they must have felt as they made their laborious "walk of shame" from one ambulance to the other, followed by the sprayer: the untouchables of the modern era. No wonder many patients ran away before they could be transported to an ETU.

We gave ourselves a short break before our return trip. The ambulance crew routinely missed lunch, but someone produced cold sodas—a welcome treat on a hot day. Elvis, who would soon be spending part of his time at the new ETU opening here in Margibi and so would not be around to watch over the ambulance crew, gave a pep talk to its members, exhorting them to follow meticulously the infection-control procedures designed for their protection.

The ambulances carrying the two patients headed back to Bong ETU earlier than usual, and we followed suit. By the time I arrived at the Bong ETU, I had been on duty for over twenty hours. I grabbed a late lunch left over in the doctors' office and looked for a ride back to my guesthouse. Tomorrow I would be back to the day shift after four straight nights; my body was scarred with mosquito bites and blisters from nightly battles with the Nairobi flies.

November 7, 2014: Time Will Heal

Apparently the ambulances had gone out again yesterday in the late afternoon after I was dropped off. They were told it would be a short ride to Gbarpolu, but it took them four hours there and another four to get back. They did not return with their five patients until after midnight.

James, the hiccup man, passed on early in the morning, not long after he asked me for water. All of his relatives lived far away and no one could come for his burial. So he came and went alone. God bless his soul.

It was a bittersweet day for Christine, the little girl who climbed out of the bed she shared with her brother, Ryan, when he died. Day twenty-two of her stay in the ETU and—her Ebola test having turned negative—she was finally going home. Her short life had been a tough one, losing both her mother and brother. She had been here almost as long as I had and had won over many hearts. In the morning, she sat on the white plastic chair at the devotional service. Deep lines of sorrow were carved into her young face. When I held her for the last time, she managed a sad little smile.

Duo, the physician's assistant who sometimes worked with me, would be heartbroken when he came in tomorrow to see her gone. I hope fervently that time would loosen and heal the harsh, dark grip imposed on her little heart by untimely death.

There were many patients in both wards this morning, some rather ill, and Vasco and I made rounds for three hours. Being encased in full PPE for this period of time did not bother me somehow. Perhaps the weather was not as hot, even if it was wretchedly humid.

A six-year-old boy named J. Godpower had started to bleed from his IV site. We controlled it by applying a pressure dressing. I was curious about his surname.

Mother Fatu's little boy, Jackson, who was Ebola negative and being cared for by Comfort in the special holding area, was now registering a fever. In the morning, Comfort brought him to the triage area and I used a rectal thermometer, with margarine as a lubricant, to confirm that his temperature was elevated. With that Jackson had bought himself a second admission into the suspected ward. Comfort would be staying with him. Indeed, all the children from the two families admitted a few days ago showed distressing symptoms.

Updated data from International Medical Corps' monitoring and evaluation team showed a reduction in mortality in the ETU, down from 64 percent a few weeks ago to 59 percent. Not much of a reduction, but a decrease nonetheless.

In the afternoon, the ambulances went to fetch three patients at a center run by the district council. By the time they arrived, one patient had expired and another had run away. The burial team took the deceased to the morgue. As the ambulance pickup pulled into the drop-off area for triage, the single remaining patient jumped out and took off, leaping into the dense jungle. The coordinator sent word to the police to watch out for him, but he had disappeared.

At the end of the day, I put on PPE again to admit a patient and move three who had tested positive into the confirmed ward. The three were Mai, Pai, and Alfred K. Pai's husband's Ebola test had not been received, so he would remain in the suspected ward. He helped us to pile Pai's belongings into two buckets; blanket, sheets, clothes, toiletries, bottles of water and ORS. Unlike most Liberians, Pai did not speak English, and she looked lost when separated from her husband. We trooped down the gravel path through the orange net fence to the confirmed ward and showed them their beds. Andrew, who had been in the confirmed ward for a spell, came and told me that Mai was his sister.

As we left the ETU after a long day, a nearly full moon shone brightly in the eastern sky.

November 8, 2014: Death of a Child

We retested Jackson and, unfortunately, he was now positive. We had failed to spare him from infection. Could we have tried harder to argue against his admission into the confirmed ward to be with his infected mother, which mostly likely sealed his fate? Meanwhile, Fatu's other child, her three-year-old daughter, Theresa, who had one of the highest Ebola titers in the county, died this morning. She lay curled up in bed, her face bloated beyond recognition. All morning long Fatu was lying in the next bed and did not realize that her child was dead. Struggling with bloody diarrhea and profound weakness, she could barely take care of herself. Jackson and Comfort were a few rooms down the hall.

At the end of morning rounds, we approached her with the psychosocial nurse and told her that her daughter had died. Fatu was sitting upright in her bed. At first there was disbelief in her eyes, then despair. Fleetingly she glanced over at the adjacent bed. Having confirmed the truth of our announcement, she was overtaken by grief, her face dissolving into expressions of pain and deep sorrow. She shed no tears, instead raising her arms as if to grasp onto something, then dropping them limply by her side. I cued the nurse to ask her gently whether she wanted to touch her child. She shook her head. At home when I saw the news with images of people afflicted by Ebola I had often been moved to tears; that was one reason I knew I had to come here. Seeing Fatu struggle with her loss this morning and feeling with her as a mother, tears again filled my eyes. It was the first time I'd cried since I'd been at Bong, and I was grateful that no one could see behind my goggles. Instead my tears mingled innocuously with my sweat.

The psychosocial nurse helped Fatu get out of her bed. She walked with heavy steps down the hall into another room, closer to her little boy, Jackson.

The profound sadness in these two families—Fatu's and Krubo's—was palpable. Krubo lost her baby, Moses, a few days ago, but she still had her two children, Peter and young Faith. All three were weak, febrile, and required IV hydration. Faith was bleeding in the morning from the IV site in her foot. She trailed blood onto the wet, chlorinated floor as she walked determinedly out of the confirmed ward to join her fellow inpatients sitting outside on the gravel.

Like a cat, the Ebola virus toyed with its captured prey, weakening it until it was too exhausted to fight back and limply surrendered.

Because of the large number of sick patients in the ETU, I stayed again for close to three hours, bathed in my own sweat. When my gloves were finally peeled off, my hands looked as if they'd been macerating for hours in hot water.

As predicted, Duo was disappointed that he was off when Christine left; he would have loved to say goodbye. The moon was full tonight, and by nightfall the confirmed ward was full as well.

November 9, 2014: Orphaned

More deaths today.

Almost 90 percent of the patients in the confirmed ward now were from a single village, Taylortown. It was about twenty minutes from the ETU by the side of the main highway, a jumble of makeshift houses made of mud bricks and corrugated sheets of tin.

Pai, the tall, strong woman who came in a few days ago, had died, and her husband, who had now been confirmed positive for Ebola, had moved to the confirmed ward.

A family of three was admitted to the suspected ward. Not surprisingly, Magbla, the one who had active symptoms, was positive. She perked up after vigorous IV hydration. Her husband and her daughter tested negative and were discharged home after they showered with 0.05 percent chlorine and then rinsed with water.

Baby Jackson was inconsolable this morning, crying loudly, bothering his very sick mother, Fatu, and demanding to suckle. Fatu, probably still recovering from the shock of her daughter's death, was sitting up in her diaper, picking at her arms and struggling to breathe. In her restlessness and confusion, she pulled out several of her IV lines. We made two more attempts to place a line in her and then gave her a dose of Lasix. But in the early afternoon she expired, leaving Jackson an orphan. He was placed facedown, limbs frog-like as he cried himself to sleep. Comfort, the nurses' aide, came in the evening to care for him.

In the early afternoon, an ambulance from Phebe Hospital delivered a man for admission. Upon arrival he jumped out of the truck and into a ravine, and from there he fled into the jungle. A quick roundup did not locate him. The sight of the ETU must have frightened him. This was the second time we lost a patient to the jungle.

Saad, a ten-year-old boy who had come in a couple of weeks ago, was finally free of Ebola, but hadn't yet left the ETU. Saad's remaining relatives lived in Lofa, a distant county from Bong, and social services had been trying to contact them to take him home. His half brother, six-year-old J. Godpower, came in gravely ill and had not turned around. Every day after finishing his breakfast, Saad came into J. Godpower's room to coax him to eat and drink. J. Godpower was too sore in the mouth—and bleeding—to want food. Saad, on the other hand, had an enormous appetite. After wolfing down his meal, he went in search of food that other patients were too ill to eat. Soon we would ask the meal service to give him two foam boxes of food for each meal.

This morning a nurse asked Saad if he would remain in the ETU to take care of his little brother. Tears promptly flowed from his eyes and down his cheeks; he stared straight ahead and said no. The place must have been hell to this ten-year-old, who was too young to bear the burden of caring for his ailing brother and being witness to more suffering and death. Also, he and his brother had different surnames. In Liberia, men can have many wives, and women sometimes have children with different men, so family ties in the villages are often a tangled web. Though it would be hard to leave his little brother, Saad was eager to return to his family.

In the hot midafternoon, I walked outside the ETU and took the winding path through the forest to the cemetery. The grave diggers were off on Sundays, so all was quiet save the chirping of a few birds. Twenty more grave markers had been stuck in the ground since I last visited, and now there were about sixty.

A plaque read:

IN LOVING MEMORY OF

ALFRED

SUNRISE: 01/01/2004 SUNSET: 31/10/2014

Fifteen more graves had been dug; the earth was still fresh.

As I lingered in the quiet cemetery wishing the departed peace, I did not feel the eerie feeling I usually have in a graveyard; instead I felt a deep sadness over all these lives prematurely ended.

November 10, 2014: Surrounded by Human Suffering
Krubo lost her second child early this morning—little Faith, who was wandering around two days ago trailing blood on the wet floor. She looked peaceful in death, as though sleeping soundly. Krubo now looked quietly depressed; her five-year-old, Peter, had stopped talking the day he arrived.

Alfred K. was restless yesterday and walked up and down the hallway, his breathing a bit troubled. He had come into the ETU with Pai and Mai. Pai died a few days ago, while Mai was struggling to stay alive across the hall from her brother Andrew, who was recovering. As I was exiting the ward, Alfred K. asked me for water. I got him a one-and-a-half-liter bottle of water, and he sat outside drinking it. The last patient I'd brought a bottle of water was James, who had died a few hours later. Today when I started work, the night shift reported that early this morning, Alfred K. got out of bed to walk out of his room, fell, and suddenly died.

After examining Watta M., I determined that she was twenty-two weeks pregnant. She started to bleed early in the morning with intermittent contractions; the fundus (the top of her uterus) had descended below her navel. Using the recently acquired ultrasound machine, I was able to detect a fetal heart rate of 130 beats per minute. As of the late afternoon when I took two physician trainees into the ward, her fetus's heart was still beating resolutely, and she had a small amount of bleeding. Infection-induced premature labor was a distinct possibility.

The ETU continued to be the training ground for many health care personnel who would be deployed to newer treatment centers. For the last four days, the physicians had been assigned to me.

Just as I happened to get out of the doffing area, I caught Saad saying his goodbyes and was able to bid him farewell. He was clutching his plastic bag of toys from the psychosocials as he imprinted his small palm on the survival wall before walking away. One of the workers told me that he had said he wanted to take me as his wife. When the worker teased that then he would have to carry me by himself the hundred miles to Lofa County, the ten-year-old replied, "No problem!" Left at the ETU was his half brother J. Godpower, still fighting for his life.

The man who had jumped out of the ambulance and escaped yesterday was found by the authorities, carrying a machete as he came out of the jungle. He was brought to the ETU in the morning, where he explained that the space-suited people had scared him off. He tested negative and was sent home.

Four generations of a family landed in the suspected ward this afternoon. Five were confirmed to have Ebola, their blood having been taken earlier in the field and tested; two were negative, including a thirteen-month-old baby and a six-year-old girl. The baby was so distressed that she cried the whole time she was in the unit. Comfort would spend the night with the young ones, and formula was rounded up for the suspected ward. (While Comfort was with them, Bendu was helping to look after Jackson.) The children were really not out of the woods, as they had been directly exposed through their contact with Ebola-afflicted relatives.

Tonight the confirmed ward was filled to the gills. In fact, a few more mattresses had to be brought in from the supply room and placed on the floor to accommodate the new patients, especially those from the family just admitted. Meanwhile, the only patient in the suspected ward was a comatose woman named Carr, who was still breathing but did not seem to be revived by hydration and antibiotics. She was brought in yesterday without a history, but Peris, my roommate, who was also a midwife, examined her and concluded that she was recently postpartum. Although she had tested positive for Ebola, we saw no reason to move her to the crowded confirmed ward, since the suspected ward was otherwise empty and she was already on the brink of death.

Last week I clocked in about eighty hours of work—almost as much as when I was an intern putting in hundred-hour workweeks.

Every day we were surrounded by human suffering, but the staff's zeal in fighting this scourge was undiminished. The meticulous care with which we were helped as we repeatedly donned and doffed made us feel loved and cared for. Each time I came out of the ward, someone invariably came forward to thank me. It made me feel as if every drop of sweat I shed was worth its weight in gold.

November 11, 2014: A Birth in the ETU

Before we started our rounds, someone called out from the confirmed ward that Watta M. had just aborted her pregnancy. We hurried off to the donning room to put on PPE. By the time we got in, Watta M. was sitting calmly at the end of her bed in her room. No baby and no sign of anything else: she was even wearing a clean sarong.

A patient yelled to get our attention and pointed out back, behind the ward. Watta M.'s room was the last one on the left-hand side of the confirmed ward overlooking the outhouses and the patients' shower rooms. We found the baby on the harsh, sloping ground next to the fence, not too far from the latrine, cord coiled loosely around its neck, placenta attached. The baby was lying on its side, headfirst down the slope, waving its right arm aimlessly in the air. It was crying and breathing well, despite such a rude entry into the world. Trailing from it were cloth wrappers stained with blood. Watta M. must have given birth to her baby in the backyard, somehow also delivered the placenta by herself, and walked away, abandoning her child.

We knew we had to cut the cord, but how? Calmly I walked back into the confirmed ward and all the way down to the other end, where the supply room was, to look for string and a pair of scissors. I found a bungee cord, but it was too slippery to be of any use. Finally I found a neon-orange shoestring. By then a midwife who had put on PPE had made her way into the backyard. Her heavy-duty gloves decreased her dexterity in tying knots on the shoestring before she finally was able to cut the cord with the scissors. She picked up the baby. The left side of its face was covered with bits of gravel that left imprints on the tender skin after being brushed off. The midwife instinctively looked between its legs: it was a baby boy. Someone came with a red hazard waste bag for the afterbirth, the rest of the stained clothing, the scissors, and remnants of shoestring. All would be brought to the morgue.

The baby was just about three of my palm lengths and as light as a feather. He had perfect little toes and fingers and looked surprisingly pink. The midwife estimated the preemie's age as about twenty-four weeks. He was covered with a towel and brought to his mom, who by then was eating breakfast. We were fully aware that the baby's chances of survival were slim. Still, we felt blessed to see a birth in the ETU.

To my surprise the midwife did not follow us into Watta M.'s room, instead heading to the doffing area without giving us any instructions. But on reflection I understood her fear; earlier on during the Ebola outbreak, many of her colleagues died from the infection after unknowingly taking care of Ebola patients without protection. It exacted a tremendous toll on their numbers.

Watta M. was lying in bed, expressionless. She did not reach out for the baby, and when we placed him next to her, she just lay there stiffly. He made feeble attempts to suckle at her nipple. I wanted to give him some dextrose, but I was told by one of the expat nurses that dextrose was no longer the fluid given to preemies. Before that I had succeeded in giving him three milliliters of sugar water. The nurse fed him formula, which I was afraid would cause a worse indigestion problem.

We taught the mother kangaroo care—holding the baby skin to skin to warm him up. Although Watta M. was reluctant to hold her baby, we strapped him to her chest while we were there and started her on a pitocin drip to slow her bleeding. Soon her uterus began to harden and she passed some clots.

Likely the premature birth was set off by Watta M.'s Ebola infection. Word was passed to us that the coordinator wanted us to send parts of the cord and placenta to the lab and to take a blood sample from the baby's heel, so we asked for heavy-duty gloves and retrieved the bag with the specimens. I threaded the cord with a syringe but was unable to withdraw any cord blood, and the placenta looked macerated. Neither the nurse nor I was willing to inflict pain on the baby to obtain blood, so we did not get any. Later in the day, when there was still talk of getting a piece of the cord and placenta to the lab, the nurse went and retrieved the bag from the morgue and managed to get pieces of tissue into a cup.

By late afternoon the baby was reported to be gasping. The nurse who went in to see her reported that Watta M. had requested that the baby be taken away. I remembered how Krubo had kept herself far away from baby Moses when he was in a coma, as though terrified to touch him. Perhaps it was a mother's defensive instinct not to become attached to a child who would not survive. Still, I could not understand Watta M.'s repulsion toward her own baby. He was her fifth and only her second son. In a modern neonatal intensive care unit, this little fighter probably would have been saved. Later, when my roommate came on for the night, she said she would try to take some blood through the cord. We all preferred to get any blood sample only after he slipped away.

Ironically, while Watta M. brought a life into the world, if only briefly, across the hall Magbla was found in her bed, stiff and cold, life snuffed out like a candle.

And Carr, the lone patient in the suspected ward, also lost her battle to Ebola shortly after daybreak.

This morning Mai was breathing abnormally—Cheyne-Stokes respiration—and her eyes were glazed over; I held her arm and said a silent prayer. She breathed her last while I was checking on her just before I left the confirmed ward. She joined Pai and Alfred K., two others admitted to the ETU on the same day who had predeceased her. Her husband, Togbah, located right across the hallway from her, had yet to take off the three layers of clothing he arrived in, including a thick jacket, despite the heat. He was weak and remained quiet.

Tewah and Majama left the ETU today, Ebola-free. Both were women in their sixties. They showered, and afterward Majama sat on the white plastic chair by the nurses' room waiting for her ride. Tewah had suffered from a distended belly and difficulty breathing for almost a week, but she responded to prednisolone once we finally found it in the pharmacy. Although she had recovered, she had also lost both a son and a daughter—Morris and Sial. Today she raised one of her arms in the air praising Jehovah for saving her; the other arm hung on to a bra she had decided not to wear—perhaps because she had never worn this strange apparel? Both Tewah and Majama had an air of triumph as they left their palm prints on the survival wall.

Our battle against Ebola raged on. In the afternoon the ambulances—both ours and Phebe's—brought in a total of thirteen patients. Our hearts sank to see so many new potential Ebola victims, and we stayed way past our shift to admit them.

George and Munyah were some of the happiest

teenagers in the world this afternoon. After several posi-
tives, their Ebola tests were finally negative. Their friend
Andrew was also Ebola-free and rejoiced with them, but
the sadness of his sister Mai's death earlier in the day cast
a shadow over his high spirits. Andrew had often asked
me to make him tea in the mornings, so today I tossed
him his warm tea in a bottle over the fence, which he
caught, beaming. George had been waiting for this day
for some time, having been disappointed that even after
his friend Saad tested negative and left he had continued
to show a low positive titer. To pass his long hours in the
ETU, George kept busy drawing with colored pencils,
posting his artworks proudly on the outside wall. As I
passed him this morning, he said in no uncertain terms
that he expected a gift from me when he left (clearly he
remembered the plastic bag of toys that Saad was given
when he was released). Munyah was ill for many days,
bleeding from the mouth, his young life hanging in the
balance. Yet he beat all the odds and walked alive out
of the ETU.

Tonight, though, my thoughts returned to the little
baby that started our long day—a miracle in the ETU.

November 12, 2014: In Loving Memory of the Baby

Finally a day off after ten straight days of work. Fortu-
itously, it fell on my birthday. I took my time to wake up
slowly. Peris was still not home from night duty. When I
opened my browser, I saw a Google doodling of a table
full of different varieties of birthday cakes with candles.
It would be the nearest thing to a birthday cake I would
taste today, and it made me very happy.

I celebrated the day by taking a run on the campus

after a two-month hiatus from running. Back in March I had suffered a stress fracture of my left second metatarsal. Despite the persistent discomfort, I had stubbornly continued to run through it. On the day of my departure for Liberia, I finally went for an X-ray, and with the diagnosis clear, I resolved to give running a break. Years ago my sports medicine doctor had advised me not to run more than three miles and only three times a week, but instead I ran four miles six times a week. After this last episode I might have to stick more closely to his advice. Nonetheless, it felt good to run freely again without pain.

Alfred, the driver, took me around Gbarnga. He was married and had a son. During the last civil war, he fled to Guinea and so never went to school. He still hoped to go, and if not, he wanted to make sure his son had that opportunity. Except for one broken tarmac road in town, all the side streets were of the dusty, red earth variety. Alfred briefly stopped at the site of an unfinished university being built by the Chinese. Huge concrete blocks rose from the dirt close to the dormitories where the Chinese had lived. They were empty now, as everyone was evacuated once the Ebola outbreak was declared.

Up a small hill surrounded by fences was "the palace," the name Alfred gave the residence of President Sirleaf when she came to pay a visit. In Alfred's opinion she was too old and should step down so that someone else could take over. We drove by the towering walls and menacing gate of the holding center, where people exposed to Ebola were quarantined and monitored for symptoms for twenty-one days. After some aimless driving, Alfred asked if I would like to visit the palace. Could a layperson do that? I asked. Evidently I had misunderstood him, as he then drove through a small village and we parked in front of a gate with high walls topped by razor wires. It turned out he had brought me to visit the Gbarnga prison.

Through a small entrance beside the towering main gate, we walked onto the prison grounds, where we washed our hands and had our temperatures taken. Guards—a few men and one woman—were sitting around casually, none carrying a weapon. Alfred announced the purpose of our visit and we signed the visitors' book. We were then led to the office of the head of prison, who was out on the grounds at the time. In the corridor we stopped briefly to look at a whiteboard on which were written the names of the prisoners, the crimes they had committed, and the length of their sentences. Outside, many young-looking prisoners were milling around in a high-fenced compound, gawking curiously whenever they noticed me.

The neatly dressed, uniformed head of prison finally appeared and I signed his book. The prison held about one hundred, he told us, mostly men but not exclusively. The women's quarters were farther down the compound, and they were allowed to move more freely. A few of the men were imprisoned for murder, some serving life sentences.

We took a brief tour along the narrow, dingy, poorly lit corridors. The prison cells were really big rooms with two rows of at least six dirty, narrow mattresses placed tightly together. There were no dressers, cupboards, closets, or other storage areas, and the prisoners hung their meager belongings on hooks on the wall. Many curious eyes peered through wire net windows; almost all the prisoners were young men in their twenties, and all seemed to have robust bodies, probably from working out. I could only imagine the sweltering heat that must build up in the night while they tried to sleep.

To one side of the prison grounds lay a vegetable garden tended by the inmates: something to look forward to for those serving long sentences. On my way out, the head of prison thanked me for visiting and said he hoped more people would hear about his facility so it could receive some much-needed aid. After this solemn visit, I asked Alfred to drive me back to Cuttington.

While I was in Gbarnga, Peris had come home after her night shift. Now she was full of news about the ETU. She reported that when she saw Watta M.'s baby, she could not believe how pink he was. She gave him about three milliliters of dextrose. In her opinion as a midwife, he would live if he could be placed in an incubator and given a continuous dextrose drip. But no one would take him because of the possibility of Ebola. She had tried drawing blood from the cord, but it was not well formed and none came. Like me, she had refused to draw blood from the baby himself. She was distraught that Watta M. still showed no interest in the baby, who would not live much longer without nutrition and warmth. Neither of us wished to pass judgment at this time, but it was hard to comprehend a mother's lack of desire to cuddle her own helpless baby.

Saad's half brother J. Godpower, age six, passed away quietly last night. His name had not helped him.

I heard from the nurses who came back from the

night shift that the ETU baby breathed his last in the early dawn. Peris must have seen him alive and not heard that he died. His life on earth was fleeting; it left hardly a ripple in the tumultuous sea of life and death in Bong County. My heart still weeps for him.

IN LOVING MEMORY OF

BABY WATTA

SUNRISE 11/11/2014, SUNSET 11/12/2014

ACCORDING TO THE latest World Health Organization update, as of November 11 a total of 14,383 Ebola cases had been reported from the three West African countries (Guinea, Liberia, and Sierra Leone) where transmission remained widespread and intense, with a total of 5,438 deaths and a mortality rate of 38 percent. Liberia had the most reported cases, 6,878, and Sierra Leone had 5,586, followed by Guinea with 1,919. Between mid-September and mid-October, peaks in the number of new cases occurred in Liberia, 509 cases, Sierra Leone, 540 cases, and Guinea, 292 cases. Investigation of localized transmission of Ebola in Mali was ongoing, while transmission was interrupted successfully in Nigeria and prevented in Senegal.

November 14, 2014: The Valley of the Shadow of Death
After my birthday, I was switched to the night shift. In the early morning when Peris returned to our shared room from her own night shift, she was feeling weighed down by all the deaths in the ETU and wanted to unload some of her distress by reviewing the stories of many of our patients.

First, it turned out that, posthumously, our one-day-old baby had tested positive for Ebola. The cord blood was also positive, though the Ebola test from the placenta was indeterminate.

The second piece of sad news was that the little boy orphan, Jackson, had died yesterday, not long after his mother's death. Peris and I felt some blame for this, since we had not kept him away from his infected mother while he was still negative for Ebola. Having no one to take care of him at that moment, we could not protest the decision, made by higher-ups. It was not to be changed until Comfort, the caretaker and an Ebola survivor, came into the picture. We tried to console ourselves that he could have been incubating a long time, since he had been breastfed. However, I doubted that Fatu was producing much breast milk; more likely the suckling was for comfort rather than nutrition. With his death, Ebola had wiped out the entire Kerkula family. May his little soul rest in peace.

The other family that came at the same time as the Kerkulas was doing somewhat better. Krubo, who had lost Moses and Faith and had only Peter left, was slowly recovering. She paid more attention to her grooming now; someone in the ward had given her a sprucing up, and her hair was all braided in cornrows tapering toward the top of her head. She did not look like someone who had lost two children: her eyes were not red or swollen from crying, and there was not even a hint of sadness in them. Peter, who had stopped talking when he arrived, now had an appetite and was mingling with the other youngsters in the ward; he too was slowly coming out of his shell.

Continuing her account of our patients, Peris also filled me in on some of their family backgrounds. Saad, she said, had lost thirteen members of his family; he was left with only his father and a brother. So much tragedy in such a short life! Watta, the mother of our deceased patient Alfred (not Watta M. with the small baby), had now lost five of her seven children to Ebola and was also left with five grandchildren to care for. She and Saad were survivors who had to pick up the pieces from destroyed family structures and go on with their shattered lives.

We talked about Joseph, who came in with his son Jeremiah. They roomed together, and for days on end father and son alternated being slightly better and then suddenly abysmal, with episodes of intractable vomiting and diarrhea. It was heart wrenching for me to watch them witness the sudden deterioration of the other, and sometimes I wondered whether they should be rooming together. Joseph struggled mightily but lost. He was practically swimming in his own watery diarrhea; no sooner had we changed him than he was leaking through his diapers and needed to be cleaned again. Before he died he requested a shower; he wanted to be clean when he left this world. With Joseph gone, Jeremiah had been moved to the last bed of the ward, which I had begun to associate with death itself. Recently he had started having intractable hiccups, and I feared he was rapidly going downhill.

Ten-year-old Gremelier had come in with a high fever but fought everyone and refused any form of care. He was so combative that no one could put in an IV line safely. He turned away fluids by mouth, using his arms and hands to fend off any kind of assistance—as though we were there to attack him. He could not lie still and was often up and about pacing here and there and nowhere. Most likely he was confused due to an electrolyte imbalance. He started to bleed from his mouth; his gums and lips became crusted with dry blood, vampire-like. Peris told me that he had died during the night.

With my roommate recounting all these deaths, we felt truly despondent for a few minutes, as if we were fighting a huge wildfire with a single extinguisher.

During my shift, at three in the morning, Nurse Paul and I went into the wards to administer IV fluids. A deafening, thunderous rain came pelting down on the tin roof; our goggles immediately fogged up from the humidity. Two of the patients were in rooms with nonworking lightbulbs; we hooked up the IV by feel and by peering through a streak in the goggles cleared by our sweat. It was grueling. Big Boy, a man in his forties who came in during the late afternoon, was unkempt, groaning, and barely conscious. His IV was not running and the light was too dim to place another IV line; with our fogged-up goggles we were almost blind. The lamps in the hallway appeared haloed, and my partner looked like a hovering white specter. I felt as though we were walking through the valley of the shadow of death.

All these deaths had to be balanced with some good news: eight-year-old Josephine and twenty-one-year-old Otis had recovered and their Ebola tests had turned negative. They were ready to go home. The night Otis came in, he was scared and very dehydrated. In the midst of getting his IV fluids he had a seizure—his eyeballs rolled back in their sockets, and his arms and legs stiffened. We called for diazepam, which soon relaxed him; his eyes began to show signs of life. Miraculously, he recovered rather quickly and soon would be going home.

Still, death does not take a holiday.

November 16, 2014: The Silence of the Night

WHO reported that the number of new Ebola cases had decreased in Lofa and Montserrado, the Liberian counties where the early cases were reported (Lofa borders on Guinea). In response, President Ellen Johnson Sirleaf had lifted the state of emergency for the Ebola crisis three days ago. But at Bong, we had no reprieve.

Our intake area stretched far beyond the villages of Bong County. For example, we also received more than 10 percent of our patients from Gbarpolu County, near the border with Sierra Leone. From here the patients had to trek three to four hours through the jungle and then cross a river via canoe to reach a health station, from which they were transported for another four hours to the Bong ETU. The very sick ones would never make such a journey; the not so sick might be severely dehydrated by the time they arrived.

The confirmed ward was now full, with thirty-one patients; in fact, a few more beds had to be squeezed into some of the bigger rooms, making it difficult and somewhat hazardous for us to move around. I kept telling myself to pay particular attention to possible obstacles. There were also thirteen patients in the suspected ward.

We found Martha A. dead early this morning; she had been agitated and delirious in the evening as she went through the final death throes of Ebola. Restlessness had also taken over her roommate Gomai, who continuously and aimlessly picked at the edges of her bedsheets; the lower half of her body was bathed in diarrhea.

The confirmed ward was so full that the staff tried to keep families together. The Juah family came from Taylortown, the shantytown not far from the ETU. They had already lost two boys. One of them was the roaming, agitated, combative Gremelier. Mama Juah shared a mattress on the cement floor with her little boy, Prince, while her husband, Stanley, slept on a bed next to them. On the other bed was Aaron, an unrelated young man who had become so ill that he was probably unaware of their presence. Stanley was strangely quiet and solemn, glancing only now and then at his wife and child. When he asked for a change of pants and all I could find was a pair of large-size brown corduroys, I cut up some long strips of sheets to use as a belt. He was not particularly pleased with my offering, and later I saw him wearing his oversize trousers with cuffs folded and a leather belt around his waist. He looked present-able enough.

Little boy Prince and his mama were spiraling down-hill, Prince bleeding from his mouth and nose and Mama with copious diarrhea. A nurse tried to clean Prince's nose and mouth, but he was fighting and kicking. We had Mama moved off the mattress so we could change the soiled sheet and wipe the mattress with chlorine. After we finished cleaning, mama Juah finally plunked down on the bed, exhausted, and IV fluids were admin-istered to all three members of the family.

Aaron, in the next bed, had a rough day with diarrhea

and an abdomen as hard as a board; we did not rouse him, as he was finally sound asleep. We had no strong analgesics other than paracetamol; morphine was still a pipe dream. Almost half of the patients now required at least a liter of IV fluid, which meant we had to wait for all fluids to be infused before we could leave the ward. We had also started to give oral potassium and bananas to correct the electrolyte imbalance caused by the diarrhea.

Across the hall was Roland, who had intractable hiccups; he was in a semicomatose state and suffered from massive diarrhea. On rounds very early this morning it took a nurse and me some time to change all the patients who were wet and soiled, and by the end of the effort we were both mentally and physically exhausted.

Those patients who were able used the bedside bucket as their commode. On admission each patient received two buckets—one for diarrhea and vomit and the other for washing. I could not imagine many Americans—who tend to be larger people—sitting on such flimsy buckets and not breaking the thing or, worse, falling off. Many patients here, however, had strong quads and were adept at squatting fully or in part. Even in this most vulnerable position, they managed to remain discreet while on these makeshift commodes. The more ambulatory patients walked outside to the back of the ward to use the latrine.

Our youngest patient, Satta, was three years old and had been here for twenty long days. She cried and grimaced when approached and never seemed able to relax. For the first week she hardly ate, subsisting on

IV and oral fluids. The psychosocial team brought her some toys, and the older adult patients tried to take care of her. Comfort was called in and slept in the same bed with her. Her Ebola titer was finally coming down and she was very likely to recover, but it had been a slow process. To date in the ETU we had no survivors among patients less than five years of age—she would be our first. (However, most patients admitted to the ETU did not know their birthdays or how old they were; instead they gave us their estimated ages.)

Like many Ebola-afflicted children, Satta would go home without her mother. Whether she would be accepted at home would depend on how much fear and stigma her remaining relatives associated with Ebola. If they declined to take her back, she would be spending an indefinite time in shelters created by Save the Children, again with strangers, along with many children who were in the same boat. One could only imagine the psychological impact at such a young and impression-able age.

Sometime in the next week, another ETU would open halfway between here and Monrovia, in Margibi County. It would be a bigger unit, with close to a hundred beds. Since we received about 40 percent of our patients from Margibi, we expected a reduction in the number of patients coming here. But for now we were full to capacity, and there was a discussion about whether to use the morgue to house confirmed patients. I thought it would be creepy to have patients sleeping there, but the IMC country director said we would only put new patients there, who would not know it had been a morgue. It was still far from a savory idea. Besides, where would we house the dead before they were buried? The director thought they could be left outside, which only made me think of the hovering crows and dead bodies in body bags being baked in the sun.

While our patients slept fitfully or struggled through their nightmares, the jungle outside the ETU was alive and teeming with noise: croaking frogs, a persistent and sometimes piercing chorus of crickets, sounds of bats echolocating. Together they seemed to remind us that while death might silence many of our patients, the creatures in the jungle would want it to be otherwise.

November 17, 2014: Screaming in Pain

Here in Bong County, Ebola continued to create havoc.

I had been on night shifts now for four nights in a row and witnessed Aaron being reduced from a talking, walking young man to an invalid lying helplessly in bed in diapers, groaning and screaming in pain with the slightest move. At first he insisted on standing up to relieve himself—and thus maintained some dignity and independence. Two nights ago he did just that, struggling out of bed with two of us at his side. His tall, thin figure loomed as he first stood, then hunched over slightly and aimed his butt at the bucket; projectile diarrhea followed. It was painful to watch him attempting to squat and ended up half standing. Last night he was lying in bed in diapers, eyes closed, burning with fever and groaning in pain when touched. He had had a rigid belly for a couple of days now; it seemed something more serious was going on besides the bowel disruption, but there was no point in further medical investigation: in his

Ebola-infected state, no hospital would even agree to take a look at him.

We started him on antibiotics, but what he really needed most was pain relief. For weeks now we had been requesting analgesics more potent than mere paracetamol. Most patients suffered horrendously from excruciating abdominal, chest, and other musculoskeletal pain, and we were unable to provide relief. Early this morning when we did our five A.M. rounds, the scream that came from Aaron when he was being changed was heartrending. Since I could not cover my ears, all I could do was to walk as far away as possible from his room.

Prince continued to bleed from both ends and had begun to be combative while Mama Juah suffered from a profusion of bloody diarrhea. The ambulance crew planned to go out to Taylortown to round up more of their sick relatives.

Kamah and Gomai, who roomed together, were like a pair of delirious grandmothers. They both had the deer-in-the-headlights expression; Kamah sat bolt upright in bed and constantly picked at her bedsheet in agitation, while Gomai lay flat in bed. Both kept up a steady, incoherent mumble.

Across the hall, Musu, in her sixties, was frail and feathery light. She was gurgling but unconscious. I gave her some Lasix to help her get rid of extra fluid, but she was reaching her end, and in the early morning she slipped away.

Watta M. had finally flushed the virus from her body. She looked very anemic and would need to take some iron pills with her when she went home—without the baby she gave birth to at the ETU.

Krubo, also bereaved on account of losing Moses and Faith, was likewise recovering, as was her son Peter. A few days ago, he developed a swollen and painful knee and a fever. We started him on antibiotics but were not sure whether this was an infection or some form of reactive arthritis. His Ebola test was now negative, and a test for malaria the night before came out negative. It was good to see mother and son sitting outdoors side by side in the evening, enjoying a Ninja Turtles movie.

November 18, 2014: A Grueling Twenty-Four-Hour Day

Fifth night in the ETU.

There was a schedule change I had not been informed of but just happened to notice on the printed schedule I found lying around. Originally I was supposed to be off last night and switch to the day shift today, but the schedule had me down for the night, my fifth. Communication here was not the greatest.

When we came on, the news from the day shift was not encouraging. During their rotation, Prince had finally passed away following a gruesome struggle with bleeding. His grandmother Juah quietly joined him, without much of a fight. The Juah family had now had four deaths in a short period of time, and rumor had it that the other villagers were becoming hostile toward all the Juahs. Their safe return to Taylortown was an open question.

The death of Carter, a woman of middle age, was sudden and unexpected. She had been relegated to a mattress on the floor close to the door, since the confirmed ward was full. On early-morning rounds the day before she had manifested no symptoms. I wondered whether she suffered a cerebral hemorrhage.

Kamah, one of my "grandmothers," who had been picking on the edges of her bedsheets unceasingly, also passed on. Four deaths in one day! It was absolutely hard to take.

Aaron hung on, but it was difficult to watch him in his comatose state, struggling for breath. Mercifully he no longer broadcast pain, but now we did not know if he was still suffering. His belly had become softer except for a distended bladder; he would need a urinary catheterization.

Sonia, an elderly lady who suffered from abdominal pain, was groaning all night long, yet she hauled her frail body determinedly to the commode and back to bed: more argument for us to get stronger analgesics.

We were seeing sicker patients, two-thirds of whom required IV fluids, making our stays in the wards longer and longer. The night before this last, I was the first one in and the last one out, remaining inside for three and a half hours, my longest yet. That won me the dubious title of still-functioning person staying longest inside the ETU wards. Although we were timed in and out of the wards while wearing full PPE, we were never told how long we were allowed to stay inside. Doctors Without Borders permitted their workers to remain inside the wards for a maximum of forty-five minutes. No time limit here, and until recently there were no clocks inside the wards by which to time ourselves (and some of the clocks that were installed at our request soon ran out of battery power). The buddy system meant we checked on each other regularly, but when push came to shove and we were needed to stay longer, we stayed.

I had wanted to go out with the ambulance crew again on my day off, but it did not work out, and since my time here was nearing an end, I decided to go this morning right after my night shift, even though I was tired. Again there were four ambulance pickups, traveling in convoy, with me and Elvis, the head of the crew, in a 4x4 van communicating via walkie-talkie. We passed by Taylortown, where the Juah family came from. Outside the Tokaka clinic in the district of Salala, we interviewed a patient who was not feeling well but had not had any Ebola contact. He had been rounded up by the district health officer, and he looked nervous, hiding on the ground behind a shed while onlookers watched curiously from across the street. He followed instructions to walk up to the back of the ambulance and get inside; meanwhile the sprayer sprayed the area where he had been sitting and the path he took. This ambulance then split off from the convoy and returned to the Bong ETU. It was a slow day for the ambulance crew: two other patients identified by the district health officer had run away before we arrived to pick them up.

We proceeded to the hundred-bed Kakata ETU of Margibi County, which was getting ready to open in a few days. It was built by the global children's rights network Save the Children and supported by several organizations. Unlike Bong, which nestled in the jungle, the area around the ETU had been cleared of trees, so the ETU sat under bright sunlight and was quite visible from the big road. We found the temperature and humidity inside the windowless white-tarped tents unbearable, even without PPE. Other than the central hallway, there was no ventilation. Patients who were not dehydrated on arrival would soon become so. The beds were already in place, and it seemed there was still a lot to do before the ETU opened. A few fans would help! Outside, some workers were planting banana trees and small bushes, but these were not tall enough to shade the area. Elvis would have to divide his time between the two ETUs once this one opened.

Our slightly downsized caravan headed back to Bong. Exhausted at the end of a twenty-four-hour day and having had no lunch, I looked forward to a quiet evening before my day shift began tomorrow.

November 19, 2014: "The Significance of the Life We Lead"

Surprisingly, Aaron was still with us, although he remained essentially unresponsive and no longer capable of groaning. He had a catheter now: we emptied seven hundred milliliters of urine from him.

There were two burials today. One was for Gomai, the anxious old woman who was Kamah's roommate. The other was for a strapping young man named Clarence. I went to Clarence's burial. Clarence's sister Oretha, who was also in the confirmed ward, cried loudly when his body was taken away. It was one of the first times I had heard crying inside the ward. Mostly there had been no wailing or weeping there, even among relatives. Clarence's short burial ceremony was attended by a big contingent of families and friends, most of them young. The psychosocial priest invoked John 16:33: "I have said these things to you, that in me you may have peace. In the world you will have tribulation. But take heart; I have overcome the world."

Since the last time I came to the cemetery almost two weeks ago, the number of graves had increased to more than eighty. At that point the grave diggers had more than ten empty graves ready, but today there was only one complete and one half dug. They were barely keeping pace.

My time at Bong was winding down, and today I was moved back to the day shift. Instead of drowsiness, I had to battle the heat and humidity of midday.

The Ebola contagion in West Africa, like the leprosy of an earlier era, subjected its victims to isolation and stigma. It took fortitude to get up every day and face the gruesome routine of seeing many patients in excruciating pain, children who were lost and lonely, and several of the afflicted in their final hours of dying. Yet my hours were so long that I had little downtime to reflect on the happenings of the day. There were moments when, instead of resting up for my night shift, I found myself lying awake and ruminating over some of the patients who were fighting hard to survive. At times when I thought about individual patients I'd spent time with I felt a profound despondency and I found my eyes moistened with tears. I wondered what was going through their minds as they lost their grip on life.

How to cope with this everyday tragedy? In the medical field, we are often taught to distance ourselves from our patients when they are approaching death in order to protect our mental well-being and thus stay competent and effective in providing care. But at Bong, it was hard to keep that distance. For one thing, we were asked here to avoid body contact with colleagues or anyone else. Twice I reached out involuntarily to shake hands, remembering only belatedly to withdraw my hand in midair. In response we devised a kind of "Ebola air hug," which was a poor substitute for the real thing. The only persons we could touch were our patients—while still separated by layers of protective gloves and clothing.

As an infectious disease consultant in the USA, I am not directly responsible for the primary care of the patients about whom I'm consulting; usually it's the primary doctor who's in charge. At Bong, each of us cared personally for every patient. They had become our charges, and the longer they were in the ETU, the harder it was to disassociate ourselves from them.

What inner peace I possessed derived from a higher power. At the end of a busy shift when all was quiet, I reflected on whatever happened during the day, and with that reflection tried to renew my strength to face another day. The ETU patients required me to remain resilient for them. For that I needed help from above. "But they that wait upon the LORD shall renew their strength; they shall mount up with wings as eagles; they shall run, and not be weary; they shall walk, and not faint." (Isaiah 40:31)

Once in a while I went to the cemetery to look over the graves and remember each life that was taken. It helped me to have a little perspective on living. Volunteering at the ETU reminded me that we are transient beings, that we ought to live life to the fullest, but in a way that has some import. In Nelson Mandela's words,

"What counts in life is not the mere fact that we have lived. It is what difference we have made to the lives of others that will determine the significance of the life we lead."

FOR CENTURIES THE disease most associated with isolation and social stigma has been leprosy, and although leper colonies have vanished in the developed world, they still hang on in many parts of Africa. Like Ebola, leprosy stigmatizes the patients afflicted with it and the people associated with them. Ebola is a deadly infection that either gets fought off or rapidly consumes its victim, while leprosy, if not treated, lingers on, leaving an infected person disfigured. But leprosy is far less contagious, gaining ground only upon repeated exposure, with an incubation period of about five years instead of two to twenty-one days. Symptoms may take up to twenty years to appear.

One hot afternoon I took the dirt path uphill toward the old leper colony of Suakoko. It took about fifteen minutes, up and then down into a small valley with a collection of mud-spattered, tin-roofed stucco houses. On my way there, I ran into three men, one of whom called himself Old Man and decided to accompany me into the leper colony, where he lived.

The settlement consisted of about five hundred inhabitants. Just outside the colony was the long schoolhouse, closed now on account of the Ebola outbreak. Until recently a schoolteacher had come here to instruct some one hundred to one hundred and fifty children from the colony. A water pump stood close by, protected by a circular wall of palm fronds. The colony, built by Liberian president William Tubman in 1955, did not have electricity.

At the entrance to the colony was a hut made from sticks and fronds whose owner sold small items. About thirty feet away lived an old lady with leprosy. Old Man called her out of her dark, one-room shed. She bent low as she came through the curtain covering the doorway. She sat on a bench made of sticks and showed me her disfigured feet. There were no sores, just stubby toes or what remained of them. Neither she nor he knew how old she was.

Farther into the compound was an open shed for communal cooking. Many villagers were farmers, and they helped one another, especially their elders who could not work. Young villagers who were not infected by leprosy could get work in the outside world, although earlier this year, some who could not find jobs had been forced to go beg along the main Monrovia-to-Gbarnga road, about two kilometers away.

In August 2014, the colony agreed to approve the Bong ETU being built nearby, but only after long discussion, as they feared being further stigmatized. However, some inhabitants had found employment and acceptance at the Bong ETU. They recognized me and were clearly pleased that I had come to visit them.

Similar to houses of most African villages I have seen, the dwellings here suffered from neglect and disrepair. Inside one was nothing beyond a bed and clothing, the latter hanging from a line or scattered across the floor and on the walls. The whole village needed a serious face-lift and a coat of paint. Many of the villagers I met were elderly women with wasted hands and feet, missing fingers and toes. One thin old lady was still trying to eke out a living by weaving baskets from palm leaves with deformed hands and stubs for fingers.

Seeing these villagers, I recalled my first leprosy patient. It was while I was a dental student assisting in a ward in St. Pancras Hospital in London: an Indian man who had a skin lesion on his leg, a patch with decreased sensation. Our attending physician was careful to call it by its proper though euphemistic name, Hansen's disease, but in Liberia everyone used the old terminology. The patient in London could look forward to a cure with state-of-the-art treatment before it caused any disability, but the inhabitants of the leper colony in Suakoko were not so fortunate.

The administrator of the colony approached me and asked me to let the world know they were getting no financial aid or medical care from the government. Many children and grandchildren were not infected but were nonetheless shunned by the larger society. They felt isolated and unwanted.

On my way back to the ETU with Old Man, a few children from the colony were busy collecting kindling. This brought back memories of myself as a young teen creating a woodpile for my mother. For an entire year, my girlfriend and I had spent the hot afternoons after school in the forest chopping wood for the cookstove. My friend had been taken out of school by her parents because she was a girl. I was lucky to be still going. The fate of the colony's children would depend above all on their obtaining a good education.

Old Man told me that he had missed his lunch, so by way of thanking him I went into the office and brought him a leftover boxed lunch that would have been discarded if not eaten.

I was glad to have carved out some time to visit the neighbors of the Bong ETU, who were generous and brave to take the risky step of accepting another highly stigmatized group into their much-neglected space, potentially exposing themselves to more social isolation.

As I waited for my ride home, I saw Old Man carrying his boxed lunch like a prized trophy, skipping happily over new lumps of fresh orange dirt, heading back to his home. The sky over the leper colony glowed a pink hue, reflecting the rays of the setting sun.

November 20, 2014: It Is Well With My Soul

Last day of rounds in the wards.

This morning I was harboring a nostalgic feeling for my first days at the ETU. I remembered how at first I felt that everything I came into contact with might give me Ebola. When I helped to clean up patients whose fluids were teeming with billions of infectious particles, I felt a real and present danger. How gleefully I sprayed chlorine solution over and around me while reciting this silent but powerful invocation: "Die Ebola, die!"

In the suspected ward, three-year-old Nowai had lost both parents to Ebola. Bendu took care of her the first night; she was quiet and followed Bendu like a pet dog. The next day, however, when Bendu was gone, Nowai started following Monique, a patient in her early twenties. Soon the two had formed a mother-daughter bond. In the afternoon both were found to be negative for Ebola. After their showers, Monique held Nowai's hand and let her sit on her lap while they waited together in the safe zone for their discharges. Unfortunately Monique was from Monrovia and Nowai, who came from a nearby village, was going to be heartbroken again when she had to say goodbye. It was going to be very hard on this little child.

Aaron finally passed away; he had fought tenaciously for several days, unwilling to let go, as though he knew he still had a life to live, that it was not yet time to leave.

Mama Juah also died, after a long bout with bloody diarrhea. She joined Grandmother Juah and her own son Prince, who died two days ago while he and she shared a mattress on the floor. Her husband, Stanley, watched his son's death, which was quite bloody, and then his wife's. He was now quiet and strangely stoic. Psychosocial nurses were afraid that he would escape over the fence into the jungle before he was officially clear of Ebola. If he tried, he would make it. Also, there was talk about his being met with hostility if he returned to his village. (Later I learned that Ebola had come to Taylortown only after Stanley's son got sick and he tried to hide it by attributing the illness to "a leopard problem." And, in fact, shortly after I finished my stint in Bong, he did disappear and no one knew where to find him.) Tragically, more Juahs were coming to the ETU with symptoms.

Krubo, however, was recovering. Looking at her, you could not tell she had lost two children and been through so much pain in the past few weeks. Her hair was well braided and her prominent cheekbones framed a proud, strong face. Peter, her five-year-old son, left a few days ago for a safe house run by Save the Children.

Quiet Togbah, who came to the ETU with his wife,

Mai, lost her almost ten days ago. For days he would not change his clothes despite the broiling heat, instead wrapping himself in layers, the outermost being a thick winter jacket. Slowly he had regained his strength. He would survive and return to his village.

I was not sure I could weather so tragic a storm with as much equanimity and courage as many of our patients had shown.

Once after a lengthy Sunday service at Epiphany Hall on the Cuttington campus, I went to fetch my lunch from the kitchen, a cement building with huge, wire mesh–covered openings for ventilation. The enormous pots were being fired up, and thick black soot from years of cooking with firewood covered the walls and ceiling. Women sat in a circle, cutting vegetables and chicken. They could not attend church but were listening instead to a radio church service. Through the wire mesh I heard the soothing sound of a familiar hymn "It is Well With My Soul." Horatio Gates Spafford wrote the lyrics of this hymn after losing four daughters in a shipwreck while crossing the Atlantic.

When peace, like a river, attendeth my way,
When sorrows like sea billows roll;
Whatever my lot, Thou hast taught me to say,
It is well, it is well with my soul.
It is well with my soul,
It is well, it is well with my soul.

III

RETURN HOME

IT IS AMAZING how in a span of six weeks, you begin to get attached to the people you work with, especially when the risk of any one of you contracting a deadly disease is frighteningly real. On my way through Cuttington in the IMC cruiser on my last morning, I heard someone call out, "Kwan Kew!" I did not catch a glimpse of the person, but I realized that I had been here long enough for the nationals to pronounce my name with confidence. That made me feel truly welcome, and sad to leave. Many asked whether I would be returning. Alas, I would not. Unlike other expats who were on an ongoing paid contract, I was not leaving for a ten-day R&R, but was heading for home.

On our way to Monrovia, we briefly visited the new Save the Children ETU in Kakata that I had seen a few days before with the ambulance crew. The opening ceremony with politicians and media coverage was scheduled for the next day, and the banana trees were in place. Inside the windowless tents, however, this ETU would still be an inferno.

Monrovia was as hot as Suakoko, but crowded. At a busy intersection, our cruiser was sideswiped by an expensive car driven by a woman in high fashion. The policeman directing traffic came over while my driver was calling the IMC office for instructions and the woman shoved an American twenty-dollar bill at him. Deeming this adequate for covering the repair, my driver tried to get back into his seat, but the police officer tried to get a cut of the payment, precipitating a brief but unpleasant shouting match.

As we drove on, my driver told me that corruption was rampant among police officers. Even teachers at the University of Liberia were known to take bribes in exchange for giving out passing or better grades for their students: their salaries were so low—and unreliably paid—that they were unable to survive otherwise. I am always disheartened when I hear of such practices in Africa.

As in other African cities I have seen, the infrastructure in Monrovia is in far from tip-top condition. I doubt that all the problems can be ascribed to the recent civil wars. The worst is an apparently permanent reliance of many African countries on foreign aid. Many seem to have no plan for being weaned off such assistance, and no one, including the humanitarian aid organizations, seems to have the right answer.

Having been in the habit of getting up early in Suakoko, I woke up the next morning before sunrise and decided to take a short run while the air was cleaner and cooler. I ended up at Brooklyn Beach (its actual name), which was surprisingly clean, with thatched huts at intervals along the beach. The waves were huge and threatening. A person—who I at first thought was hustling me—approached to say that normally he would have charged me a fee for entering the beach.

Later, I was driven to visit the ELWA3 (Eternal Love Winning Africa) Ebola treatment unit run by Doctors Without Borders. However, we were not allowed to enter the 120-bed facility because of an incident not long ago when a visitor inadvertently wandered into a red zone. At the peak of the Ebola crisis, patients were often turned away from this ETU. Then we drove to a smaller treatment center, the ELWA2 Ministry of Health ETU, where we waited in the green zone until we decided it was not worth the wait after all.

The day was getting hotter. We went to West Point, a slum on a peninsula that stuck out into the Atlantic Ocean. Here some seventy-five thousand impoverished inhabitants are squeezed into several thousand tiny, dilapidated houses. Narrow tarmac roads and alleys crisscross the slum, making it inaccessible to our van. Squalid unsanitary conditions in the area had allowed the easy spread of Ebola, and at the end of August President Sirleaf attempted to enforce a twenty-one-day quarantine in West Point but had to abandon it after ten days. There are few toilets, and residents use the beach to defecate; many also use it to earn their livelihood, fishing. Children congregate in the front of the houses, playing on dirty cement porches next to a busy street. I am sure tuberculosis is rampant.

Back in the hotel, the news on Al Jazeera was nothing but reports of ongoing wars in Syria and Iraq, the killing of non-Muslims on a bus in north Kenya near the Somalian border, killings in northwest Nigeria, mass rape in Darfur, plague in Madagascar. There seemed no end to the senseless violence, which threatened to overshadow our infinitesimally small attempt to make the world a little safer from Ebola. I truly did not miss this kind of news while at Bong.

Two drivers from IMC accompanied me late that night as I said my farewell to Liberia. Monrovia was under curfew, and as we traveled to the airport, we were stopped at two checkpoints and asked our reason for being on the road. At the airport parking lot guard booth, one of my drivers was refused entry, and after much shouting and argument he had to get out of the van. We finally said a quick goodbye at the airport check-in, where everyone had to wash hands and look over a fact sheet about Ebola.

At the first stop on the return journey at Casablanca airport, I, along with other passengers, was greeted by personnel in PPE who took our temperatures long-distance with a machine aimed at each one of us. No such precautions were taken at our next transit point, Frankfurt.

My initial itinerary back to Boston had been through JFK in New York, but I rerouted through Washington Dulles on the off chance New York might adopt some crazy form of quarantine. After all, back in October, New York governor Andrew Cuomo had initially joined New Jersey's Governor Christie in calling for a strict mandatory quarantine of returning Ebola volunteers. He later softened New York's policy when New Jersey suffered negative publicity over the forced quarantine of

Maine volunteer nurse Kaci Hickox in an isolation tent, despite her having no symptoms. We volunteers heard all about these events in Suakoko, and after that none of us wanted to go through Newark Airport. We also remained wary about what might happen to us when we arrived at the other major airports. The public was split on how returning volunteers should be treated: as heroic citizens who had performed selfless acts or as pariahs to be forcibly quarantined in the interest of protecting the general population.

Long before I left for Liberia, I had planned not to go back to work for twenty-one days (the incubation period for Ebola to rear its ugly head after one has been exposed to it) after my return. I was told at that time there were no such rules for quarantine, but I knew that policy was fluid. In any case, it was best to allay anxiety at the hospital, even if a returning volunteer without symptoms represented at worst a theoretical possibility of infection. Every day on the front lines, ordinary people were risking their lives helping patients sickened with the virus. A blanket quarantine of all returned volunteers would only discourage future volunteers to West Africa, where confronting the outbreak would do the most in the long run to protect the global population from Ebola.

Knowing that a possibly long screening process at Dulles might make me miss my connecting flight to Boston, I had decided in advance to spend a night in Arlington, Virginia. Upon hearing that I had arrived from Liberia, the immigration officer immediately alerted an officer in a face shield, mask, and gloves to take me for further questioning. I was told to surrender my luggage claim tag to another officer, who would get my suitcase, and to follow the first officer.

I was taken to a waiting room. After an attendant came and handed me my suitcase, the officer with the mask signaled me to step into his cubicle. He reached over his desk with an infrared thermometer and took my temperature, making sure he was as far from me as physically possible. My temperature was not elevated. He seemed to become nervous when he learned that I had been in contact with Ebola patients and that I had visited a laboratory for Ebola. He asked me why. I told him I had been volunteering in Liberia.

"Are you a nurse?"

"No, I am a doctor."

Standing up and maintaining his distance, he asked me to take all my belongings and follow him, without offering to help with my suitcase. Fortunately, I always make it a point to pack lightly for my volunteer trips. We walked through another open office and through a metal door into an isolation cell with no windows other than a small glass one in the door. Both door and door window could only be opened from the outside. In the cell were a metal bench, a stainless steel sink, and a toilet. The room was cold, poorly ventilated, and stank as if someone had just used the toilet. He left me there without saying how long I would have to wait. The door that could be opened only from the outside slammed shut.

I rolled my suitcase to the metal bench and sat down. Thinking I might have a long wait, I pulled out a book to read. Soon the door opened and I was joined by a young man with rumpled blond hair, without baggage.

He walked to the far corner of the room, pacing the floor and periodically running his fingers through his hair, agitated. At first I thought he must be another volunteer from West Africa, and I asked as much. No, he told me: he was not an American but a Norwegian. Then he muttered something about having had a small run-in with the law. Suddenly it dawned on me that I was in a cell with a person who had broken the law, that I was being treated like a person who had committed a crime. The ironies were multiple and not amusing— my cell mate had been placed in a closed room with a "possibly contagious" returned Ebola volunteer without being given any of the protective gear worn by the officer who had brought me here. After a long journey from Africa, I had been locked in a room alone with a strange man and with no avenue of escape or means of soliciting help were I to be attacked. As I sat there trying to read my book, it slowly dawned on me that I might really be quarantined by the powers that be, and that I might be unable to stop it.

After about fifteen minutes, the metal door swung open. The same officer who kept his distance took me to an adjacent open room where two officers in plastic gowns, face shields, masks, gloves, and booties and a third only with a mask questioned me further about any possible exposure to Ebola. They were from the CDC. Did I take care of Ebola patients? Had I touched any dead bodies, been exposed to the blood of patients with Ebola, attended any burials? My answers were all in the affirmative, though qualified by the addition that I had always been dressed in full PPE—of a better quality and more effectively fitted than what they were wearing.

One of them was slightly apologetic about the whole process and went on to say that she very much appreciated my having spent volunteer time in West Africa. The whole screening process took about an hour, though it certainly seemed a lot longer. I probably would have missed my flight to Boston if I had booked to fly out the same evening.

At the end of the screening, I was given a big manila envelope and told that someone from the Massachusetts Department of Public Health would contact me about its quarantine procedure. Finally, before I could go, I had to wait until they had cleared with the state of Virginia that I would be spending the night there. Could I use the restroom, I asked at that point. The officer who kept his distance flashed a worried look and pointed to a restroom in the far corner of the airport. Was he afraid it would have to be sanitized after I used it?

Finally, once I was cleared to leave, I heaved on my backpack and rolled my suitcase to the exit, liberated for the next leg of my trip. That I was free now seemed quite arbitrary, given the precautions taken inside the airport. Although some states had prohibited "Ebola returnees" from traveling on public transportation, here I was, ready to share the hotel bus with passengers who had not been told of my potential quarantined status, let alone offered protective equipment. In the morning I would be allowed to board a commercial aircraft from Washington to Boston. When exactly was my quarantine period supposed to start?

Shortly after I returned home, the Massachusetts Department of Public Health contacted me. According to them, my twenty-one-day quarantine entailed direct surveillance: I would have to report my temperature and symptoms twice a day, and someone from the health department would have to lay eyes on me in person or via Skype once a day. I was not allowed to go into crowds or where people congregated. Technically I was not quarantined, as I would live at home and could still go out for a run, a walk, or a quick trip to the store—I could always pick an hour when they weren't busy. *And what about the poor uninformed cashier getting exposed to me?* I thought. DPH knew about my kitten. The official informed me that if I did get sick, she would probably have to be euthanized. Did I know, he asked, how much it had cost the state of Texas to quarantine the dog of the nurse in Dallas who had contracted Ebola while caring for Thomas Duncan, the first Ebola patient in the US?

On the morning of my second day at home, I received a call from DPH informing me that three officials, two from the state and one from a local health department, would be arriving shortly. Rather than drawing attention by knocking on my front door, they wanted to meet me at the back door—but they would not enter my home. I figured that since they told me they had decided not to wear hazmat suits, the neighbors would just presume they were Jehovah's Witnesses.

When my doorbell rang, my kitten, Grisela, immediately ran to the back door. I scooped her up. She was curious and watched the three health department personnel with her big eyes as they stood on my back porch, asking if I had any symptoms and politely inspecting me—from a distinct distance, as if I had grown two horns. I was the first person in the state under quarantine, so I was of special interest.

That afternoon, the photographer and reporter from the local *Belmont Citizen-Herald* who came to interview me were happy to sit with me and Grisela on my living room couch for an hour. The reporter's father, a cardiologist, had assured her that she need not be concerned about the visit.

On my second day back in Belmont, I received an invitation to view the seasonal decorations at the White House—a follow-up to the volunteer service award and pin I had been given in the spring of 2014 in recognition of my past volunteer work. The White House event would take place on December 6—an ideal birthday present for Tim, my oldest child, as that happened to be his birthday. Unfortunately it would also be only the sixteenth day of my twenty-one-day quarantine. When I inquired with the Massachusetts DPH about going, they reminded me I could not board a commercial vehicle or be in a crowd and suggested I check to see if I might go on another day. However, they also asked the Centers for Disease Control, whose opinion was that it would be all right for me to attend. There was still a hitch, though: I could not travel in a commercial vehicle and would need permission to travel through the states of New York and New Jersey. In the end I had no choice but to skip the December 6 White House invitation.

While my husband came to welcome me back at the airport, it would be another five days before Charles, my youngest child, would return home for Thanksgiving, which we were all looking forward to. Although we did not really talk about it, it was clear that they would all be with me during my quarantine period. They had their respective jobs that they would go back to, which involved being in close contact with students. If I were to come down with Ebola, the ramifications would be huge. My husband, who is a professor at Boston College Law School, and my son, who taught public school in New York, would certainly create a headache for the governors and the Departments of Public Health of the respective states.

Overall, although everyone seemed to commend me on my volunteering, the actual homecoming had been less than altogether welcoming, except for my family and kitten. It's not as if this surprised me. Donald Trump tweeted on August 2, 2014 that "The U.S. cannot allow EBOLA infected people back. People that go to far away places to help out are great but must suffer the consequences!" I was well aware that Americans readily succumb to media-induced panic: the BBC had even reported that the main reason for a shortage of protective equipment in the African nations actually afflicted with Ebola was that the equipment was being hoarded in the US for use in a theoretical outbreak here. Before I left for Liberia, I borrowed two books from my local Belmont library asking for a lengthier time than the usual four weeks, scarcely daring to breathe a word about bringing the books to Ebola Land. Even then, afraid to be branded a pariah when I returned, I only told the few neighbors I met by chance about my going to Liberia.

About a week before the end of my quarantine, the Boston papers reported that someone who had recently traveled from West Africa had been admitted to Massachusetts General Hospital for possible Ebola infection. I wondered whether some of the people who knew where I'd been thought that person might be me. Sure enough, I received an email inquiry from a reporter asking me if that was so.

In fact, I had had two moments of possible exposure during my days at Bong. The second came a few days before the end. While I was doffing and my goggles were already off, I felt fluid flowing down from my hood to my left eye and burning it. That day the sprayer was spraying me more vigorously than usual, including over my hood. I could not reach out to wipe my face and just withstood the burning, trying to reassure myself that the burning was probably from the chlorine and not from any infected fluid. Still, a lingering—and probably healthy—fear lurked in my mind.

I have seen deaths in my medical career but never so many in such a short period of time. It was as though death was mocking our impotence. In this daily tug-of-war we would typically win a few hours' reprieve with our IV fluids and medicines, then lose it all again a few hours later. Most terminal patients died a lonely and scary death, unattended, struggling with their last breath. Children separated from their parents almost certainly could not comprehend why they had been abandoned. But we also relished the sporadic moments of victory when our patients managed to spurn death's sting. Taking care of Ebola patients in the ETU was not only distressing emotionally, it was also physically challenging, especially while wrapped in full PPE. I was glad to have kept myself in good enough condition to withstand these conditions without becoming demoralized.

Those of us who came voluntarily to help in the

outbreak have been called both heroes and villains. I am certainly no hero, and I cringe at the thought of being labeled a saint. I am just an ordinary human being who responded to our fellow humans' cry for help, nothing more. There is no question in my mind that this was the right thing to do despite the risks, including the possibility of becoming infected by and dying from Ebola. *Sed Ministrare.*

THE DAYS OF quarantine actually passed rather quickly. The first week happened to be Thanksgiving week, so I just pampered myself, resting and reveling in the loving folds of my family. However, at night, I began to wonder anxiously whether I would be able to deal with the sudden emptiness of the house once my son left again and whether the memories of Suakoko would overwhelm me emotionally. I remembered that for several months after I returned from volunteering in Libya during the Arab Spring, the sound of a helicopter always made me think I was back in the midst of war.

During my second week, I ran, took walks, enjoyed Belmont's Habitat Educational Center, read, wrote, cleaned the yard, and raked the leaves from the old copper beech. While I raked, Grisela would hide in the piles of leaves and wait patiently for a suitable opportunity to sabotage me. I kept myself busy, and before long it was time for Christmas decorating. I was even reinvited to view the White House decorations, and a few days before Christmas my oldest child, Tim, and I went to Washington to do just that.

Although I was basically home free once my quarantine was over, many in West Africa still faced a much more uncertain period of quarantine. In Liberia, patients who were Ebola-free carried around with them certificates issued at the ETU as proof of their immune status. Back home, administrative officials at my hospital asked for certification that it was definitely safe for me to participate in public events and care for patients, so I asked the head of communicable disease control for the Massachusetts DPH, Dr. Alfred DeMaria, to write them that it was okay for me to return to work. Fortunately the coverage of my stay by local papers did not spread undue fear or make me feel I had to carry around a printout of Dr. DeMaria's email.

Appearing in public after a period of quarantine seemed a little surreal. Some people avoided me, some stood at a distance when greeting me, some approached me tentatively but without extending their hands—and a few gave me heartfelt hugs, which I appreciated. I no longer presented any threat, but people would have to find their own comfort level by themselves and on their own time. Full acceptance might not come until the Ebola epidemic had long passed.

Not long after the beginning of my quarantine, I received a message from International Medical Corps asking me about a possible redeployment. I knew that I wanted to go back to Africa, but my memories were still raw and painful and I was not ready to do so right away. So I waited for half a day before answering. I was willing to go immediately if they truly needed me, but it turned out that a mid-January date was also fine. We agreed on that, and I also expressed my wish to go to Sierra Leone this time. I wanted to go there as a doctor treating patients and not doing administrative chores, so I turned down the offer to become a medical coordinator.

A week or two later, a human resources employee from IMC called to ask if I would be interested in going to Sierra Leone as a director of training. I was firm in my wish to go there as a clinician and not in a more supervisory role, but I agreed to be flexible about training personnel as part of my responsibility. I realized that if the number of cases continued its slow decrease, fewer volunteers would be needed inside treatment centers. Training local personnel in contact tracing and prompt isolation would be the key to halting the outbreak altogether, given that one remaining case could spark an outbreak.

Not surprisingly, many people were amazed that I wanted to return to West Africa, and I was peppered with questions. To mention the most important, and my responses:

Why did I wish to go a second time? Didn't I count myself lucky to escape the first time? Although I remembered two distinct occasions when I might have been exposed to contaminated fluids during my stay at Bong, the busyness of the ETU routine and the daily selfless acts of my colleagues had alleviated that fear, and soon I did not even count the number of days after exposure.

Would my presence in Sierra Leone make a difference? Sierra Leone had the worst ratio of health care workers to population of the three battleground countries and had suffered the highest casualty rate among that meager cadre. More physicians were needed on the ground, where loneliness was pervasive and palpable. Even if all I could offer at an ETU was care and compassion through the layers of a moon suit, that was better than nothing, and might ease the ache in the lives of a few.

Was I deterred by the quarantine? The quarantine was an inconvenience but probably necessary to allay unfounded public anxiety and fear. Although I sometimes felt shunned during and after my quarantine, these feelings paled in comparison to what many Sierra Leoneans must have felt when they had their quarantine period rolled back to day one every time a relative came down with the infection. Some had been quarantined for as long as ninety days, all the while running a danger of contracting Ebola from a loved one.

While the news on Ebola was being replaced in the US media by more sensational events—and new cases had decreased dramatically in West Africa—Sierra Leone continued to have the highest number of cases and a fatality rate that kept on climbing. I also had a compelling personal reason to pick Sierra Leone as my country for redeployment. A few years earlier, Charles, my youngest child, had spent twenty-seven months as a Peace Corps teacher in the Port Loko district, where Ebola was now rampant. I was proud of him for taking the courageous step of living and working in an unknown country under harsh conditions in order to help total strangers. Ever since, he has retained a fondness for the country and its people. It seemed fitting for the mother to follow in her son's footsteps.

*Kwan Kew Lai in front of sign to Bong Ebola Treatment Unit in
Suokoko District, Liberia.*

The entrance to the Bong ETU.

Doffing: being sprayed with chlorine before doffing.

Lai just after doffing.

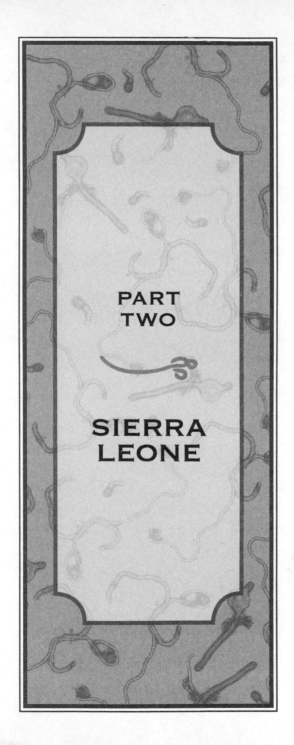

PART
TWO

SIERRA
LEONE

IV

ARRIVAL AND SETTLING IN

ON MAY 24, 2014, following a miscarriage, a woman in eastern Sierra Leone was admitted to Kenema Government Hospital. Because of the recent Ebola outbreak in a nearby area of Guinea, a health care worker tested her for Ebola and placed her in the isolation unit established there to handle cases of Lassa fever. The woman turned out to be positive and became the first confirmed case of Ebola in Sierra Leone. WHO was immediately notified by the Ministry of Health. The woman survived, and since appropriate precautions had been taken, no patients or health care workers at the hospital became infected.

That kind of luck did not last. Contact tracing pointed to the source of the woman's infection as a traditional healer from a small village in Kailahun District, east of Kenema. Her miraculous powers were so well-known that many Ebola patients from Guinea came in hopes of being healed by her. Unsurprisingly, the healer contracted Ebola and died. Because of her reputation, the healer's funeral and burial were attended by hundreds. Investigation by local health authorities suggested that the deaths of 365 people could be linked to their having participated in the funeral of this healer, during which it was common cultural practice for family and community to touch and wash the body of the deceased in preparation for burial. Thus, despite the survival of the first diagnosed patient, the outbreak in Sierra Leone soon blossomed. Before long it led to the deaths of twelve nurses.

Kenema Government Hospital has the only Lassa fever isolation ward in the world, and its contact-tracing staff quickly became overwhelmed as the epidemic swelled both there and in the neighboring Kailahun District. Sierra Leone and its neighbors already suffered from a serious shortage of health care workers (on average only one to two doctors for every hundred thousand people); now a significant number of these few were among the dead. At the end of July, Dr. Sheik Humarr Khan, head of the Lassa fever department at the Kenema hospital and one of the world's experts in viral hemorrhagic fever, died of Ebola at age thirty-nine. By mid-September of 2014, 318 health care workers had contracted Ebola, 151 of them dying from it. As in Guinea and Liberia, the virus spread to the country's capital, Freetown. The overcrowded conditions and the fluidity of the population in this city made it an ideal milieu for fanning the infection.

During the last quarter of 2014, several NGOs built treatment centers for Ebola in Sierra Leone. I was going to the one in Lunsar run by International Medical Corps, which had opened at the beginning of December. No destination had been mentioned when I was asked to be redeployed, and it did not matter to me, although Lunsar was in the same Port Loko district where my son Charles had taught some years earlier. Besides Lunsar, there was talk of sending me to the larger city of Makeni or perhaps to some screening unit.

I was originally slated to leave for Sierra Leone on January 19, but as so often happens when planning a trip with an NGO, my actual departure date did not arrive until several weeks later. I did not regret the delay, since it allowed me to experience the Boston blizzard of 2015: sixty-one inches of snow in less than thirty days! Despite all the shoveling, I loved the white stuff, and my kitten was amazed and confused by the vast crystalline expanse. Meanwhile I tried to prepare myself mentally to switch gears for the heat and humidity in Sierra Leone.

On February 10, I left for Freetown by way of Brussels and two layovers in West Africa: Dakar and Conakry. Sierra Leone's Lungi Airport is located on a peninsula that requires a twenty-five-minute ride by sea coach or speedboat to Freetown. By this time, evening had fallen. A strong, brisk wind roiled the ocean water, and the unlit floating dock bounced severely with the surf. On the dark beach, children played and some asked for money by gesturing with their hands to their mouths. Once ensconced in the departing speedboat, we swayed wildly upon the waves. I noticed that the life jacket I put on was missing a zipper and tie strings. For a few minutes, some fifteen passengers bounced violently around on the water while far away the lights of Freetown glinted at us. Even as the boat became steadier, I could not help but think of the watery grave that would be ours should the boat capsize. As we got closer to the dock in Freetown, another speedboat whipped past us through the water, without any lights.

At the Freetown dock, a fellow volunteer doctor and I were met by another IMC driver, who took us to the Family Kingdom Resort, a plush tourist accommodation on the waterfront in the Aberdeen district whose grounds were home to a couple of dik-diks and a nyala.

In the morning I walked across from the hotel to Lumley Beach, took off my flip-flops and went for a run. Back at the hotel, a man showed me a headline on the local newspaper that read, "733 Quarantined in Aberdeen." Later that same day we learned that the local authorities had cordoned off one of the nearby beach sections after some new cases of Ebola had cropped up, apparently brought by a fisherman—a fact that made me regret my barefoot run.

After a brief orientation at the IMC office, which was not far from the resort, we drove on a surprisingly smooth road for two hours to Lunsar, fifty miles north of Freetown. We passed stretches with vibrant markets, small villages, oil palms, and coconut, mango, and guava trees interspersed with tall elephant grass. Here and there we saw random tracts where bushes and trees had been slashed and burned.

All the buildings in the area of Lunsar where we would be staying seemed to be coated with a fine red dust. It had been dry, and the trucks coming to and from the iron ore mines constantly churned up tiny particles from the red dirt roads. The dust coated everything: houses, trees, vehicles, clothes on clotheslines, and an otherwise inviting stream. At the sign for

Bai-Suba Resort, we turned on to an unpaved road, closely flanked on one side by a cement-block wall and on the other by an unfinished building. We passed through a village of cement or mud-brick houses, many apparently unfinished; fields planted in cassava, potatoes, and okra; and piles of uncollected plastic garbage and bottles next to houses with men, women, and children outside milling around. Farther on, a thick brown stream was pooling into a big mud puddle where a truck and a car were getting a wash, and children bathed and women washed clothes in the brown, murky water. Eventually we reached an iron gate marked Bai-Suba Resort.

My room was spacious, with a queen-size bed—and air-conditioning and Wi-Fi once the generator was turned on at around six in the evening. After we deposited our bags, we drove to the IMC office, about ten minutes down the main road, and met with the training director. I wondered whether he was filling one of the positions I had been offered. Strangely, he had not worked one day inside an Ebola ward. Nevertheless, he would supervise our three-day training, a requirement before I could start working inside the facility.

Bai-Suba had been built by a wealthy government official to be a meeting place for owners and businessmen connected to the mines. There were some fifteen clusters of single-story buildings, each containing two to four attached rooms. The place looked newly built and was devoid of trees: huge stumps could be seen here and there on the grounds. The compound was expansive and totally enclosed by a cement-block fence topped by barbed wire. The rooms were quite luxurious, with individual bathrooms and a front porch enclosed by low cement walls and an iron gate.

Alas, sleep was difficult to come by. My room was located close to the gigantic generators that provided air-conditioning but also rumbled loudly all night long. Besides that, the mattress I was sleeping on was a hard box spring. I felt like the princess on top of the pea, tossing and turning all night long. Perhaps I was also energized because of some news from home. Before I began my day, I had received an email from my family with a link to the National Public Radio piece about my Ebola experience in Liberia ("The Ebola Diaries: Trying to Heal Patients You Can't Touch"), which had been broadcast that very morning. Their global health correspondent Nurith Aizenman had spent a whole day at my home in December interviewing me. Beyond that, Nurith and NPR had devoted serious time in researching and preparing the final broadcast, and I was very impressed. The positive reactions I received as a result left me feeling very humbled.

TRAINING BEGAN LATE the next morning for me and another new doctor who was to become the assistant training director. As we walked into the treatment center, our temperatures were taken; we washed our hands in 0.05 percent chlorine and had the bottoms of our shoes sprayed, then walked into a room where we changed into scrubs and boots. The boots were neatly shelved by sizes, but the scrubs, though stacked in cabinets marked for size, were in fact placed randomly. Unfortunately they were also mostly large, engulfing me in unwanted cloth. As we stepped out of the changing area, we were greeted by a sign in both English and Krio: "Welcome to Lunsar ETC." (Ebola treatment facilities in Sierra Leone were called centers, not units, as in Liberia, perhaps because this was the term favored by the British authorities who administered them here; thus the acronym ETC instead of ETU.)

Since the Lunsar ETC had opened only at the beginning of December, it had the advantage of learning from other units around West Africa. Based on the blueprint of an MSF installation, it had the same number of beds as Bong—fifty-two—but the layout was less congested and, as at Margibi in Liberia, the tents were made of heavy-duty white plastic instead of blue tarpaulin. The high ceiling provided a lot of breathing room, though I rather missed the blueness and intimacy of Bong.

There were two tents for the psychosocial group next to a tent for changing into and out of scrubs. The unmarked and unsheltered gravel paths of Bong were replaced by concrete walkways under a tin roof. On the right side was the WASH (water, sanitation, and hygiene) tent, followed by the medical tent where sign-out happened and health care personnel gathered; on the other side were both the open triaging area where the ambulances brought in patients to be admitted, and the donning and doffing areas. There were in fact two doffing areas, with four doffing stations. All areas, including the donning area, were partially open, allowing for ventilation by light breezes. Getting into our gear in such an open area was cooler and thus less hurried. On this first day of orientation, our program consisted only of a dry run of donning and doffing (which had some minor differences from Bong, such as wearing two pairs of gloves instead of three), and a tour of the facilities.

Beyond these areas for triage, admission, assembly, and changing were the wards. Here the ETC had three units: suspected, probable, and confirmed wards. The probable ward housed mainly patients displaying more symptoms of Ebola—especially "wet symptoms" such as diarrhea, vomiting, and bleeding—whereas the suspected ward kept patients who had a low probability of Ebola.

Past the wards down the cement walkway were the laundry room and an impressively organized, spacious, air-conditioned pharmacy—and yes, it did have morphine and tramadol! Here was also a kitchen where food was cooked for patients and staff and a separate dining area (unlike Bong, where we had to do practically everything in the congested doctors' office). Behind the long wards were visitors' tents, while farther up to one side of the confirmed ward was the morgue tent. The incinerator was an ancient, rusty structure that resembled a squat tin man and was located close to the morgue. It had not been busy of late. Next to it were open pits for the burning of regular waste.

After a walk through the grounds, the trainer took us to look at the new training center still being constructed outside the ETC itself. Evidently it was meant to provide ongoing training programs. The classrooms and dormitories were already up, but not yet the mock ETC.

The white tents of the ETC were sweatboxes that lacked both fans and windows for cross ventilation. The very sick patients who stayed inside sweat profusely, making hydration all the more urgent. This was a big design flaw, given the hot and humid climate. At the medical tent, the temperature would also get hot by noon, and by sign-out at two in the afternoon, everyone was invariably sweating and thirsty. On this first day, we walked around in full PPE for an hour and a half, but we were not in the hot zone and the weather was a kind seventy-seven degrees Fahrenheit.

In the Lunsar ETC, doctors were given six-hour day shifts, either eight A.M. to two P.M. or two to eight P.M., with the night shift running a full twelve hours from eight P.M. to eight A.M. The national nurses had the same schedule as the doctors and so, originally, did the expat nonphysicians. Sometime in January, I was told, the American nurses became bored during off hours and, stuck in Bai-Suba with nothing to do, and with the generators turned off during the day, turning the rooms into infernos and cutting off Wi-Fi, they began to come back to the ETC to help out.

Soon the expat nurses started taking twelve-hour shifts, night or day, just as they did at the Bong ETU.

As part of our orientation, we were briefed on the patient census. On our first orientation day, we arrived after sign-out or handover, but we learned from one of the expat doctors that there were currently only eight patients, evenly split between the suspected and confirmed wards, with the probable ward empty. Later that afternoon we watched the discharge of three young patients: Nyandewah, Sheka, and Georgieta. One by one they emerged from the shower, weak and visibly shaken from their days of ordeal. First came twelve-year-old Nyandewah, thin and unsteady on her legs but greeted by the psychosocial team singing, dancing, and drumming. Then came ten-year-old Sheka, wearing an oversize T-shirt that came down below his knees; he was serious, unsmiling, and did not appear joyful at all. Finally there was Georgieta, seventeen, thin but steadier than the first two. After a while I saw a faint smile flicker across Nyandewah's face. The staff gathered to celebrate with them and everyone marched down the covered walkway to another tent near the exit of the ETC, where, after they had lunch, the psychosocial staff provided counsel and handed out their Ebola-free certificates. One of the expat doctors gave Sheka a ball, but he remained sad. I wondered what ran through his little head. All three would be returned to their villages to live with relatives, as their parents were gone. It was a bittersweet moment.

Besides the discharged trio, a two-months-pregnant woman who had recovered from Ebola was being transferred to an MSF-operated center to get an abortion. According to one nurse, the patient was not happy about the abortion, but her husband thought it was the safest alternative for their family given the other children in the house. No newborns had ever survived Ebola, and MSF would not keep her for the duration of her pregnancy. This difficult choice brought back to me the memory of Watta M.'s premature baby, born in the backyard of the Bong ETU, who lived for only one day.

Our second day at the Lunsar ETU fell on Valentine's Day, and when the trainer put us through another dry run on donning and doffing, I had a Valentine's heart painted on the back of my hood. Afterward we had to go for a mandatory lecture on sexual exploitation and abuse, even though I had heard the exact same presentation online last year.

In the late afternoon a doctor, a national nurse, and I donned and drew red hearts on our hoods and went into the suspected ward, where since yesterday's discharges only one patient remained. The ward was luxurious compared to Bong's: both spacious and brightly lit. The patients even had bedside tables and real IV poles instead of the wood posts with long rusty nails that we made do with at Bong. The hall was twice as wide as ours had been and lined with at least twelve stations for hand washing (at Bong we started with only two).

The afternoon was hot, and the lone patient had moved his mattress outside, so the ward itself was completely vacant: what we saw was just empty beds spaced far apart. Each of the several clocks had stopped at a different time, depending on battery life. The patient outside was feverish, very thin, and coughing. Although he was Ebola negative, he had a history of tuberculosis. He did not remember when he started tuberculosis treatment but knew he had not taken his medications since August, when St. John of God Catholic Hospital in Lunsar was closed by the government on account of Ebola-related deaths. The tubercular patient said he had not been tested for HIV; he would be transferred to Port Loko Hospital for further treatment once his second Ebola test returned negative.

The confirmed ward was too hot for all of its occupants save one. This was thirty-five-year-old Claudetta, who was ailing and weak but not deathly ill. The rest of the patients were lying on mattresses outside; before long Claudetta wearily joined them. Sulay, a twelve-year-old boy who had been here for three days, was weak with a high fever, diarrhea, and abdominal pain. We asked him to drink a cup of ORS, and a minute later he promptly brought it all up. His older cousin Samuel was a young man, also quite stable. Zainab, a young woman, was suffering from diarrhea again after many days without it. Since there were so few patients, we were able to make our rounds more slowly and methodically, pausing to examine each patient. Besides the nurses and doctors, these patients were tended to by a caretaker who had recovered from Ebola. Her main job was to encourage them to drink their ORS while we were gone.

Ebola cases had been decreasing in Sierra Leone until recently, when on February 11 there was a bump, with seventy-six new cases reported. The resurgence came both in the Port Loko district where we were and, as we saw the day after we arrived, in Freetown.

I left the ETC as the sun was setting among the palm trees. The dusty, dry landscape looked like a desert.

As part of my three-day training, I participated in a morning session with the center's psychosocial workers. As in Liberia, the psychosocial team here dealt with issues pertaining to family members affected by Ebola such as separation, death and loss, placement of survivors without family, and care of orphaned children. They gave us a quick rundown on the current social hierarchy in Sierra Leone: local political power in the country still ran along tribal lines and depended on ruling families; each district was divided into chiefdoms. The paramount chiefs were the sole authority over the chief in each chiefdom, which usually comprised a few villages. Each village in turn had a village chief followed in diminishing authority by a mama queen and a youth leader. The psychosocial team worked with the chiefs to get patients accepted back into their villages, since the villagers tended to listen to the chiefs.

As the outbreak dragged on, the stigma originally attached to a single Ebola-infected patient was transferred to entire Ebola-quarantined villages or areas. Ebola survivors once shunned by their own villages were eventually welcomed back because they were supported by humanitarian organizations with food, clothing, bedding, and a monthly stipend of 500,000 leones ($60-$70). As it is the custom of Sierra Leoneans to share what they receive, some relatives stood to benefit from such homecomings. However, the returning patients continued to grapple with a loss of livelihood, since few people wanted to buy produce from Ebola survivors.

A quarantined village was guarded by soldiers or police. Food was delivered by humanitarian organizations and pushed under a barrier into the cordoned area. Sometimes the delivery did not come in a timely manner and so the psychosocial team would have to advocate for the villagers with the appropriate authority or organization. And whenever one person in a quarantined village came down with Ebola, the rest had to restart their quarantine period from day one.

Sierra Leoneans treat their sick ones with a lot of care and love. The sick person becomes like a child—he is washed and fed, made to lie down on someone's lap while his head and body are stroked and touched. A dead person has to be washed and touched before the burial. Ebola had changed all that. There was to be no

washing or touching. The Lunsar ETC had no graveyard, so patients' bodies were taken away to be buried
either in Lunsar or Port Loko cemeteries. If a patient
came from very far away, relatives had to make do
with the distant burial site. If they could not attend the
burial, they were given the plot number so they could
visit on a future date. Hearing this made me appreciate
all the more the proximity of the gravesite at the Bong
ETU, and the sense of emotional closure provided by a
visit there.

Despite progress in many areas, fear and distrust of
authority continued to be pervasive throughout Sierra
Leone. According to the head of the psychosocial team,
as recently as two weeks ago in a village not far from
the nearby city of Makeni (Bombali District), a couple
secretly relocated their sick child to a separate hut. They
gave her little attention or food, and she died a few
days later. When officials came to the village for a head
count, they recruited another girl as a stand-in. Fearing
that they would be found out by the village chief and
fined, the couple tried to throw their daughter's body
into a latrine in the dead of night, but it proved too
small. In desperation, they contrived to dump her body
into a well but were caught.

Another issue discussed at our session was the difficulty of enforcing sexual abstinence upon recovered
Ebola patients, especially men. Despite counseling to
abstain from sex for ninety days after being Ebola-free,
a young man had sex with his wife, who contracted
Ebola. He was being held on a charge of intent on an act
of grievous bodily harm. At the time of the outbreak, no
one knew how long Ebola could survive in semen (we

now know that the virus can remain in semen for up to a year or longer, even if the blood is virus-free). So the best methods were abstinence or using condoms.

In the late morning after the psychosocial session, I put on my PPE and went into the wards with two other doctors, one from Sierra Leone and the other from Pakistan. Two new patients had been admitted to the suspected ward: Foday, a middle-aged man, and Mamusu, a young lady. Her main problem was weakness and a reluctance to eat and drink; we urged the caretaker to prod her in our absence. Ebola tests for the two were sent to Port Loko, half an hour away, though it often took up to a day to get the results back.

In the confirmed ward, Claudetta's neck was like a bullfrog's—swollen lymph nodes made her unable to open her mouth wide enough for us to see her pharynx. We started her on antibiotics—ceftriaxone and clindamycin—but her fever persisted. Because of her pain, I was leaning diagnostically more toward a bacterial than a tubercular lymphadenitis.

We finished rounds in less than an hour. During my morning psychosocial lecture, the other doctors had walked around the perimeter of the ETC in their scrubs to check on the stable patients by querying them over the fence. The sicker ones in bed inside could be attended to in more detail later. It was a smart way to help curtail the time spent in the wards in full PPE—the maximum allowable time in the wards in full PPE was an hour and a half.

After rounds I was shown the procedure for charting patients. Signs and symptoms were recorded every shift in the wards. In the morning these were called out over the orange fence to a scribe, who transferred the information onto another form. Since anything brought into the wards was considered contaminated and could not be taken out, this procedure allowed us to have the information on hand outside to help us make further clinical decisions. No such consistent charting existed outside the wards at Bong: all our charting was done and stayed inside the wards and could not be reviewed at leisure. Also, finishing documentation and ordering while inside the wards had made our rounds that much longer.

Now that I had been through the wards twice, I was to be put on the schedule with the rest of the doctors starting the next day. I had been assigned the afternoon shift, two to eight P.M.

On the morning of my first regular workday, I went to the market in Lunsar with another doctor who would be on duty with me in the afternoon. The market had been closed Sunday afternoon, so on this Monday morning it was very crowded. As in most of Africa, the streets had no sidewalks and we were forced to walk in the road, motorcycles and cars honking constantly. A loudspeaker blared earsplitting music. Given the ABC (avoid body contact) edict, we carefully considered whether we really ought to plunge into the crowds. The other doctor had almost finished her time here, so she led the way. We tried to wriggle around, avoiding skin contact, but it was almost inevitable. The market had fresh produce, fresh and dried fish, bread, spices, palm oil, kola nuts . . . and as always, no shortage of Chinese-made goods, such as basins, buckets, slippers, and radios, as well as secondhand clothes from America.

Young girls carried aloft basins with fresh mint, potato leaves, and onions to sell, while the boys followed us around, one of them asking for money. Schools remained closed on account of the Ebola outbreak, so many school-age children were out and about selling stuff. St. Peter Claver Catholic Church, with its quiet facade, stood guard at the end of one of the streets. It was closed, so we could not take a peek. I had in mind to pay it a visit one Sunday.

V

DOING ROUNDS

ON MY FIRST day at the ETC, it was overcast all day and a cool breeze blew across the tented halls as we put on our protective gear. Except for the fogging of the goggles, getting outfitted in this unusually pleasant weather gave us no trouble at all.

The first few days of rounds, we seemed to have between six to eight patients in the wards, with admissions coming in dribs and drabs. For the health care personnel, it was both agreeable and a bit frustrating that many of the patients being admitted were not suffering from Ebola at all. The virulent, contagious nature of the disease made it imperative that the admission process be more inclusive than exclusive. The decision to admit was based on an algorithm set up by International Medical Corps. Anyone who had a fever and had had contact with an infected patient was admitted. No fever but general bleeding or a history of contact with a person who had died suddenly was also grounds for admission. Contact but no fever required the patient to have at least three symptoms before being admitted.

One example of a false admission was the case of Momoh, who was lying on a mattress outside on the gravel when I first saw him. He had not moved his bowels for six days, and his abdomen was so distended that he looked six months pregnant. He had not been in circumstances likely to expose him to Ebola, and an examination showed that the distension might be an incarcerated hernia. Both the Sierra Leonean doctor and I tried to reduce the hernia, without success. We wanted to transfer him as soon as possible to Port Loko Hospital for possible surgical intervention, and mercifully his Ebola test result turned out negative. Unfortunately, a patient declared suspect still had to have two negative Ebola tests two days apart before being discharged. Luckily, we were able to get an exception to the rule and move Momoh to a hospital in Freetown, where an Italian surgeon could perform the necessary surgery.

As with any group in confined quarters, the Lunsar wards developed their own codes and rituals. Deborah's grandmother Mabinty could not take big gulps of ORS, so we told her, "Drink slowly, slowly, or take small sips." In the Temne language, this was *moon atonaton atonaton*, a rhythmically repetitive prescript that Dr. Kamara repeated in a singsong manner, inducing delighted smiles from some patients in the confirmed ward. It soon became the morning mantra for the rest of the patients.

Dr. Kamara had completed his internship at the government hospital in Freetown. The previous August several of his friends and colleagues had died of Ebola. According to him, it was more dangerous to work in the government hospital than in a controlled place such as our treatment center. After his friends died, he went to work for three months at a German-sponsored holding center located in a police training station. Dr. Kamara's parents and older siblings had tried to talk him out of working with Ebola patients, but he felt he could not stand back and hide while his colleagues were getting sick and dying, whatever his own fears. His parents and siblings stopped communicating with him and he was forbidden to use the front door to enter the house or to use the sitting room. In November he joined the International Medical Corps and came to Lunsar; he was happy working here, and his family had since become more accepting of his work, as no one working with him at the ETC had become sick.

ON FEBRUARY 20, my day off, five of us went to Bureh Beach, on the Atlantic just outside Freetown, about a two-hour drive. During our expedition I learned that one of our own volunteers, after having a fever for a few days, had found her way to the ETC for health care workers in Kerry Town, not far from Freetown, home to a treatment center that had a dedicated ward for health care workers who became ill with Ebola. All of us who worked at the Lunsar ETC had our temperatures checked by the drivers before any trip and at work every day before we entered. It was hard for my colleague to face going into a treatment center as a patient and experiencing what every one of our patients goes through.

She was the only patient during the first night, and on subsequent days two more patients joined her. For a time there was talk of a medevac to bring her to the

US. Days in the ETC became lonely and anxiously long with no more interaction than a few phone calls and no access to the internet to distract her. It was an ordeal having to wait for the results of two Ebola tests. Her first was negative, but while waiting to hear about the second, she was informed that the medical team was putting on PPE to come into the ward and speak to her. Immediately she assumed this meant bad news, since a negative result could have been shouted out across the barrier when she was outside the ward. For some time her anxiety built up, until the team was fitted out and came inside to tell her she had tested negative a second time. Their reason for donning PPE could just have been that they wanted to deliver the happy news in person.

Leaving brought its own anxieties. In her haste on arrival, she had made the mistake of bringing her passport and yellow fever certificate into the ETC in a Ziploc bag, and for an anxious moment she was asked to surrender these to be bleached before discharge; bleaching would have most likely destroyed them. After much argument, she was allowed to save them from that fate. She was slated to head home soon to the USA—but would the CDC, knowing she had been in the Kerry Town ETC, demand that she be quarantined in Africa for twenty-one days before being allowed back?

THE WORLD HEALTH Organization announced in mid-February that the total weekly case incidence of Ebola had increased for the second consecutive week, with 144 new confirmed cases reported in the three countries during the second week of February. In Sierra

Leone, there were seventy-six new confirmed cases, with a resurgence in cases in the western district of Port Loko continuing for a second week, despite the mobilization of many community social workers trained to spot possible Ebola cases for prompt medical response. Here in Lunsar, however, things were surprisingly calm, and it often felt as if we were in the eye of the storm. Lunsar had not had new cases of Ebola since January. Most of the new cases came from the other parts of Port Loko District. The volunteers who had been here longest observed that the town of Lunsar appeared more bustling than when they first arrived. They wondered if the townspeople had become less vigilant about close contact and gatherings.

The Makeni treatment center in Bombali District, on the other hand, was filled with new patients from a village right outside Makeni. Apparently an Ebola-infected person had escaped from the quarantine in Aberdeen, the fishing community in Freetown, and traveled by minivan to Makeni, infecting a number of people on the way. (In Sierra Leone's packed public minivans, it is virtually impossible to avoid body contact. Once when Charles and I were traveling from Gbinty to Port Loko in such a vehicle, a carsick baby behind us threw up on Charles's back!) Apparently the infected man, having somehow escaped detection at the Ebola checkpoints, had gone back to his village of five hundred to seek the help of a traditional faith healer. The healer's treatment included touching and washing him. The man died twenty-four hours later, and within a week his extended family, father, mother, uncle, and brother, as well as the traditional faith healer, all ended up at the Makeni ETC. The man's father and the faith healer died within two days.

Makeni had only five patients that day but soon registered some fifty new cases, with thirty-one confirmed. On February 27 the Makeni ETC confirmed seven deaths, and by then the army had placed the village on lockdown. The flare-up, coming right after the report of two new cases in Sierra Leone, was a vivid reminder that it took only a single symptomatic Ebola patient to spread the infection. There was talk of transferring some patients to Lunsar ETC since we remained relatively quiet, but it seemed the Bombali authorities preferred keeping their patients within the district. Instead, one of our doctors and some national staff were sent there to help.

The news from Bombali was grim, but sometimes reports of outbreaks in the outlying areas failed to generate the tidal wave of admissions we feared. On February 21 we were advised that another ten patients—relatives of the four confirmed patients in the ETC—would probably be brought in from the Marampa chiefdom area. By the afternoon of the next day, only one of the ten admissions had materialized, with two others that might still show. Although we had nine patients in the probable and suspected wards, by the end of my shift the following day, most turned out to be Ebola negative. Overall, except for a spike in the third week of the month, our February census remained fairly steady at between six and eight—a great relief to me after having experienced the crowded Bong ETU.

Some patient cases were truly mystifying. Take Ramata—a thirteen-year-old girl brought in by an ambulance, comatose, with no information. The nurse coordinator called her mother, who said that Ramata's sickness was the work of the devil. She seemed afraid to be in the same space with her daughter and gave no more information about Ramata's condition. We undid the girl's sarong-like wrapper, or *lappa*, to reveal a chest with delicate breast buds; her hair was all braided in cornrows, tapering at the top into a pointy tuft. She groaned feebly when an IV line was put in, but I could not elicit much response with sternal pressure or painful stimuli. Her neck was stiff; she was jaundiced and had an enormous spleen. We started her on IV artesu- nate for cerebral malaria and ceftriaxone for bacterial meningitis, but she was dead by the middle of the next afternoon. Her Ebola test was negative, but even so her mother believed her sickness was the work of the devil and would not visit.

Over the course of my workdays, I got better acquainted with my Temne-speaking nurse, Asiatu. She explained to me that she had been working at St. John of God Hospital in Lunsar until it closed down a few months earlier, after several health care workers died from Ebola. As a single parent without income for several months, she found it difficult to care for her three children, aged two to sixteen. She told me that she knew of women who had resorted to prostitution to bring in money. She and her children lived with her mother, who ran a small business that provided some support. When she first started at the ETC in November, she lied to her mother, saying she was not taking care of patients. Asiatu's mother was the fourth wife of a husband who had six wives and a total of twenty-seven children.

Some portraits of Lunsar patients:

SAMUEL & SULAY

This young man and his twelve-year-old cousin were admitted together a couple of days before I arrived while displaying many of the usual signs of infection: fever, diarrhea, and a wrenching pain in the gut. As soon as they tested positive, they were moved from the probable to the confirmed ward. The boy Sulay was the more volatile, having trouble keeping down his ORS. Luckily this gastrointestinal problem cleared up within a few days, and though he remained weak, he was quickly on the rebound. Samuel, on the other hand, never displayed much in the way of distressing symptoms, even as the infection lingered within.

After my first day of rounds, the cousins were the only two patients remaining in the confirmed ward, and they spent most of their time outside, steadily improving. Sulay asked regularly for oranges and ice cream, whereas Samuel's treat was a raw cassava (manioc or yuca) tuber. Unprocessed cassava contains cyanogenic glycosides that get converted into poisonous hydrogen cyanide when chewed. However, the kitchen staff did not blink an eye when we conveyed his request. Actually, I myself ate quite a bit of uncooked cassava root when I was young—thankfully with no ill effects. We assumed Samuel knew what he was doing. Indeed, eating that raw cassava did not hurt him—perhaps he was immune to the cyanide. Anyway, Ebola seemed by far the deadlier poison.

Sulay and Samuel got better every day, slowly regaining strength and even beginning to show a sparkle in their eyes. They were tested daily to see if the virus had been cleared, and Sulay's test turned negative about a week after he arrived. Samuel's test, however, was still positive. The tests were repeated with the same result, and Samuel argued with the staff that if Sulay were discharged he should be as well, since he was the only relative left to take care of him. Despite outward signs of recovery, Samuel did not receive his first negative test until almost a week later. His younger cousin had celebrated his liberation from the treatment facility three days earlier and lived in a safe house until his cousin was also free and available to serve as the boy's caretaker. Samuel finally left the ETC two weeks after he entered it.

NNAFAT

Ten-year-old Nnafat lived less than a week after arriving at the Lunsar ETC. On February 21 she was brought in by ambulance from a quarantined home in Lunsar, where she had been living for some time—kept apart from her mother by the ban on visits. Admitted right away to the probable ward because she had vomited in the ambulance, she probably could not comprehend why she had been dropped here by herself, completely alone. Separated from her mother and everything else familiar to her, she would cry, "*Ya!*" ("Mommy" in Temne) when someone touched her. Whenever the caretaker tried to clean her mouth, she pulled away. My attempt to examine her was greeted with tightly shut eyes and the exclamation, "*Gbe pe me!*"—"Leave me alone!"—an echo of the harsh reality that her mother had, even if unwillingly, abandoned her.

On the fifth day, Nnafat started to have bloody diarrhea, the sign of a more severe infection. She was brought outside, where she lay on her side on the mattress. The bleeding from her mouth that had begun on day four continued unabated, and she had a wound on her left foot that was also bleeding through its dressing. She closed her puffy eyes tightly and refused to open them— or to eat or drink ORS. All she would accept was a bit of plain water. She died two days later.

DEBORAH & FATMATA K. & FAMILY

Shortly after afternoon handover on February 16, the ambulance delivered two patients from a quarantined area in the town of Lunsar: Deborah, age sixteen, and her cousin Fatmata K., four; both had been exposed to Deborah's mother, who had died of Ebola on arrival at the ETC in early February. Fatmata K. was sick with a fever and diarrhea. Deborah was totally without symptoms. Nonetheless, we decided that both should be admitted, together, into the suspected ward. Deborah's head was covered by a black hijab studded with silver ornamentation, and she was dressed in a long, sinuous, embroidered outfit that she would have to part with before she stepped into the wards. Fatmata K. had been mute from birth. Later, when we donned PPE to see her in the ward, she became terrified to see us standing beside her bed all suited up; worse still, she had to have an IV line placed in her arm so we could give her artesunate for the treatment of malaria (our protocol was to treat anyone who came in with a fever for malaria whether or not they had been tested for it); instinctively she reached for her cousin in the next bed.

The next day the tests for both Deborah and Fatmata K. came in positive. Deborah remained asymptomatic, so her result surprised us. We probably should not have been, since when she came in she had reported that there was another sick person living with her and her cousin in the same crowded, quarantined area. Fatmata K. was still weak as well as terrified of us. Deborah, who had been worried about her four-year-old cousin, was moved with her into the confirmed ward.

Two days after her admission, the sickest person in the confirmed ward was Fatmata K. She still refused to eat but could be coaxed into drinking some water. She refused to drink ORS, so we continued her IV fluids. Deborah, meanwhile, had started to turn febrile and feel muscle pain. That same afternoon, while I was on rounds, the ambulance brought in Mabinty, Deborah and Fatmata K.'s grandmother, and Kadie, Fatmata K.'s mother and Deborah's aunt. They both lived in the same household within the quarantined area of Lunsar where Deborah's mother had contracted Ebola. Mabinty was sixty-four, tremulous, but did not have a fever, whereas Kadie had a high fever and diarrhea. Getting up from a sitting position posed an enormous problem for Kadie: first she had to lean her body weight onto a table, then push herself up, and even so her legs seemed to give way from under her. She was wearing a long dress that hid her legs, and we could not clarify what kind of disability she had. The aides handed her her cane, and she almost fell while taking a few steps on her own. In the end we decided to transfer her to the probable ward on a stretcher.

Kadie's high fever and Mabinty's diarrhea continued.

They looked weak, with Kadie more deflated and exhausted than the older Mabinty. Dr. Kamara and I were accompanied on rounds by Aminata; Dr. Kamara spoke Krio and Aminata spoke Temne. Our advice this day to Mabinty to go slow in drinking her ORS was the occasion for the coining of the ETC slogan *moon atonaton atonaton*. Mabinty did not develop many wet symptoms over the following forty-eight hours, and she briefly reported feeling stronger. Kadie seemed sicker, and her disabled right leg made her illness that much harder to handle. Somehow she contrived to use the bucket by half sitting on her bed, but in the process often spilled her diarrheal stool on the cement floor. Both she and Mabinty did not want the morning visit we arranged for Kadie's four-year-old daughter, Fatmata K. When staff brought her to their bedsides, Fatmata K.'s mother just shook her head and waved her away without so much as a word of endearment. The caretaker then turned to Grandmother Mabinty in the next bed, but she also shook her head and did not reach out. Poor, suffering Fatmata K. was carried away by her caretaker with crying eyes and an open, voiceless mouth. As much as I could understand the difficult predicament of mother and grandmother, I could not help but feel that a brief, tender embrace and some loving words would have gone a long way.

Indeed, Fatmata K. was quickly going downhill. Her breathing had become rapid, with mild upper abdominal retractions and indrawing, and when she was upset she showed nasal flaring. Perhaps luckily, that very night she expired—finally released from her lonely and agonizing sojourn in the ward. While she was alive, the four-year-old had seemed to be always grimacing in pain or profoundly afraid, crying without tears or uttering a noise. In death, no longer tortured by her inability to communicate with anyone, she looked peaceful. Her own mother and grandmother had been unavailable. Only her cousin Deborah showed concern for her. I wondered how the three survivors would cope with her death.

In fact, Mabinty suddenly became delirious and confused and passed away the next afternoon. It happened just minutes before handover, and one of us physicians had to pronounce her dead before the burial team could be called in. Meanwhile, Kadie turned critically ill—still conscious, but complaining of painful swallowing and continued profuse diarrhea, with her eyeballs rolling up to reveal the whites of her eyes. Another Temne-speaking nurse, Sulaimen, said Kadie was trying to die. While her aunt was losing steam, Deborah seemed to fight back with every ounce of her energy. Whenever I asked her how she was doing, the slender and delicate Deborah would flutter her thick black eyelashes and say "fine," no matter how bad her diarrhea and abdominal pain. All of us were pulling for her.

Two days after admission, Aunt Kadie was gasping for breath, her breathing reduced to shallow and short sighs, head thrown back, mouth open and crusted with old blood. Her eyes were only partially open and had lost all luster. Her kidneys shut down and she stopped producing urine. Her death a few hours later, not long before our shift ended, left Deborah as the only remaining member of the family.

Alas, she too had taken a turn for the worse. Normally she enjoyed lying on her mattress outside to escape the inferno inside the ward, but on the day of Kadie's death we found her lying on her bed, stripped naked except for her diaper, her caretaker sponging her carefully in an effort to lower her temperature. She was awake, and I could see traces of blood in her nostrils. For the first time ever she looked disarmed, and while she looked at me I felt absolutely helpless.

Nevertheless she told me bravely that she was fine. Over the span of three days, three members of Deborah's family had died in the shared ward. So much tragedy in such a short time had failed to kill her positive attitude; her eyes were still bright and lively, and I prayed that somehow she would pull through. When I next saw her, she had a high fever and the caretaker had again stripped her down to her diaper and was sponging her. Her nose had stopped bleeding, but the diarrhea persisted and she complained of abdominal pain that was only slightly alleviated by paracetamol. I wrote her a prescription for tramadol, a stronger pain medicine. About that same time, the burial team arrived at the morgue to fetch Kadie's body for its final journey to the Lunsar Cemetery.

Another day passed and Deborah, who was sharing a room with Nnafat, also quite ill, continued to struggle bravely. Luckily, each girl had her own caretaker—a real luxury compared to the help situation at Bong. Deborah could still summon the strength to walk on her own power. True, she was weaker, was eating only small amounts, and became exhausted very quickly. She continued to insist to me that she was fine, but she dozed off as I was speaking to her.

Back in Lunsar at dinner on my day off, I learned that although Nnafat had died in the afternoon, Deborah, the only patient in the confirmed ward, was still holding her own. In fact, her condition varied widely. Much of that night, she lapsed into respiratory distress, but by morning she was able to lie with her head only slightly elevated and not feel short of breath. Still, she had been weakened so much that she could no longer walk without assistance. Thankfully she was not actively bleeding and her diarrhea had stopped. She repeated her mantra: "I am fine." When I asked her what she would like to eat, she requested couscous and soft drinks. If she did manage to eat couscous and drink, I thought maybe that would be the sign of a turning point.

That night, however, the diarrhea turned bloody and she began hearing voices. She reported to the nurse that around her bed she had seen four of her dead relatives— her mother, grandmother, aunt, and cousin—calling for her. Within a few hours, we had lost her.

Looking back, I often wondered if we could have done more to save her, yet in the end I think we did all we could. During her entire ordeal, this brave young lady kept on saying she was fine. I was working at St. John of God Hospital in Lunsar when she passed away. I would have liked to attend her burial, but my schedule and miscommunication made that impossible.

MAYENI

Around noon a week after I started doing rounds, a thirty-eight-year-old woman named Mayeni walked up to one of the gates of the Lunsar ETC. Alerted to her arrival, I walked out the long gravel path with my Temne translator Favour in order to question her. We found a diminutive woman squatting on top of her handbag next to the iron gate, a *lappa* covering her head and shading her body from the fierce sun. She told us that she had been sick for three days, despite having been seen three days ago at the Lunsar community clinic. She claimed to have had no contact with the Ebola virus or with a dead person; nevertheless she had a myriad of symptoms (fever, headache, weakness, abdominal pain), and these bought her an admission into the suspected ward. When we asked her for a contact number, she reached underneath her dress and pulled out a dirty brown bag, fumbled in it, and handed us a piece of folded paper with a number written on it.

Not long after, a nurse in partial PPE walked with Mayeni along the gravel path outside the ETC and into the ward, followed by a sprayer. Seeing this parade of three strangely clad people on a long and laborious walk under the hot sun reminded me of the leper who walked into the city to seek healing from Jesus, all the while obligated to cry to the crowd, "Unclean, unclean, make way for the leper."

Three days later Mayeni had entirely recovered from her fever and tested negative for Ebola. Everything she had brought with her into the facility, including her brown bag, was burned except for the few items that could withstand thirty minutes of soaking in 0.5 percent bleach. Wearing a pink T-shirt with a Chinese character that means happiness, she was all smiles as she headed for the shower to get ready for home.

VI

FROM THE ETC TO ST. JOHN OF GOD AND BACK

FROM THE OUTSET of my discussions with International Medical Corps about a return trip to West Africa, the organization had wanted to use me for training or screening. Although my heart lay in helping patients one on one inside the treatment center and I had refused to take on any administrative duties, I had agreed to assist with screening. So, starting on the first of March, I began periodic day shifts at St. John of God Catholic Hospital in the town of Lunsar. Also known as Mabessaneh, the hospital is over fifty years old. In August, however, it had to be closed temporarily on account of Ebola-related deaths. In September, sixty-nine-year-old Brother Manuel García Viejo, the hospital's medical director, contracted Ebola and left the country for Spain, where he died. In January the hospital had been reorganized by several expat doctors as part of a collaborative effort by hospital personnel, the Lunsar ETC, and health care workers from the African Union. Besides the screening operation, the outpatient department had been reestablished and was open six days a week. The inpatient unit had only just started up and was yet to be fully staffed; many patients were still afraid to come anywhere close to the hospital.

My first day there, a Sunday, I learned that on Sundays the hospital was closed, though staff continued to see emergency cases that required triaging. So I stayed on and had plenty of time for another doctor to introduce me to hospital procedures.

The triaging area was carefully partitioned into safe areas and red zones, with patient and staff flow carefully directed so that traffic was always heading from clean to contaminated areas. Staff wore either scrubs, partial, or full PPE, according to where they were working. Separate hot-zone rooms were designated for pregnant women, and for patients with "wet" or "dry" conditions, all very organized and carefully thought out. The patients deemed good candidates for the ETC were shepherded through a separate gate, where the ambulance came to transport them. Of course, there were also donning and doffing areas. As always, the bottoms of our shoes had to be sprayed with 0.5 percent chlorine, which we also used for washing gloved hands; bare hands we washed with the less caustic 0.05 percent solution. In the triaging rooms where patients entered, staff wore partial PPE while taking temperatures, while sprayers matter-of-factly coated the chairs and patient paths.

To me the ETC was probably the safest place for us health care providers, as we were always in full PPE when caring for patients. In the triaging rooms at St. John's, on the other hand, any patient who walked in to be screened could be an Ebola patient, and we were not always wearing any protective gear (although it's true we triaged across a wide window, with patients in a separate room). The doctors in the outpatient department of St. John had to have a lot of faith in our ability to rule out possible Ebola patients. If we made a mistake, they might be exposed to infected patients—though, thankfully, the outpatient staff also regularly wore some protective gear. On the flip side, we could make the opposite mistake and send a non-Ebola patient to the ETC, where they would mingle with Ebola-infected individuals while waiting for their test results and thus be inadvertently exposed. This was especially problematic in the Lunsar ETC, since patients in the suspected ward shared rooms (unlike the Bong ETU, where patients had private rooms and were discouraged from mingling). Altogether, it was a dicey situation.

That first day at St. John's was remarkable for its inactivity. We were told that earlier that day the fledgling inpatient department had performed a C-section on a pregnant woman who had arrived with her baby dead in utero. Otherwise all was quiet on the hospital grounds, except for a few dogs cavorting after taking long naps in the cool dirt. Outdoors near the hospital canteen, several women were cooking a delicious concoction of rice with onions and peppers in a curry sauce in several big pots. One of them told me that for twenty thousand Le I could have lunch and a drink whenever I came to St. John. Then, just before midafternoon, we had a patient: a young man who came in from Makeni via ambulance with fractures in his leg.

I had been told that on weekdays the triage at St. John of God was much busier, with around thirty to forty patients a day. Indeed, when I showed up again at the hospital a few days later, at least twenty patients were already waiting patiently in chairs set up for them outside the hospital entrance, some of them quite well dressed. One woman in particular stood out in her bright-fuchsia traditional clothes with a headdress to match; she was sitting bolt upright on a stool with a wide smile. Many of the others were women with young children. "Wet" patients, with diarrhea or vomiting, would have been triaged first (and the area decontaminated), but that day we had none. Throughout the morning into the early afternoon, we triaged a total of thirty-nine patients—none of them suspicious cases for Ebola. They were all sent farther along to see the outpatient doctors.

At the beginning of January, on the day the triaging area first opened, the outpatient unit had seen six patients; gradually, as word spread, the number had increased. Meanwhile the inpatient ward still held only a handful of patients, all in the pediatric unit. It would take time for it to reach full capacity.

One common theme that reminded me of the importance of St. John's reopening was the premature termination of tuberculosis treatment caused by the Ebola crisis. Patients who had been diagnosed with TB before the outbreak often found themselves unable to obtain their medications, since the health care facilities they had used had closed down. I personally saw more than one patient who had gone without medication for some time, and I imagined that a similar situation existed for patients with HIV/AIDS.

During the afternoon of my first busy day at St. John, while I was visiting the pediatric ward to check on the babies, I heard wailing from one of the rooms. A young woman was sitting on the floor against the wall, crying loudly and wiping her eyes with her *lappa*. She was inconsolable. Her four-month-old baby had just passed away, probably from pneumonia, but because of the ongoing Ebola outbreak the mother was not allowed to hug or touch the baby while the hospital staff alerted the burial team. The baby was lying alone on a bed behind a screen. When the mother reluctantly dragged herself from her baby, still holding on to a corner of her *lappa*, she was choking with sorrow as a river of tears flowed down her cheeks.

Before long I discovered that Saturdays as well as Sundays were slow at St. John's—probably because the outpatient department was only open for half a day. No matter: the slow pace gave us a chance to experience the strangely soporific town life during the Ebola scare. Stores closed by noon because of the epidemic. Women gathered under the shade of trees to cook, wash clothes, and style their hair, some with bare chests on account of the heat. The men congregated in their own groups, around vehicles and shops. Goats roamed around, sampling small sprigs from short bushes. One man gave a huge sigh and said wearily that Ebola had destroyed everyone's livelihood, and he was tired of it.

As we waited for patients to arrive, a big group of traditionally dressed men in boubou or *agbada*, women in boubou and headpieces, and section chiefs in white traditional clothes, suddenly arrived at the triaging area, all distinctive and elegant. These, it turned out, were the community leaders, including the paramount chief, identified by the pendant medal he was wearing.

They were being given a tour as part of the hospital administration's effort to publicize the reopening of St. John's. Later I spotted them roaming the outpatient and inpatient areas.

Our relatively free Saturday gave me an opportunity to check on the very first patient we had seen on our first day at St. John. This was a baby brought in by a very worried couple. Though the mother had been unable to tell us the baby's age, we guessed three months: she had been having profuse diarrhea for two days and was refusing to breastfeed. She'd had no fever or vomiting, and her parents denied she had been exposed to Ebola. Her sunken dry eyes, desiccated mucous membranes, and limp body had betrayed severe dehydration. We'd asked mom to try breastfeeding again, but the baby was too weak to do so.

We had trudged farther into the hospital to look for a doctor to administer intravenous fluids but were told by an attendant that the doctor was in a meeting and could not be disturbed. Once more I felt the fatalism that often influences the reactions of Africans to urgent situations—the sense that nothing can be done and one must let the inevitable take its course. Unwilling to be defeated, we'd asked the attendant if she knew of any other doctors who might be free; she told us to find the doctor ourselves. My partner, a Sierra Leonean national, had pleaded with her to tell us where we might find one, since we did not work regularly at St. John's and would not recognize a doctor if we met one. Fortunately at that moment we spotted some official-looking people standing near the entrance to the outpatient department and went over to them. Two foreign doctors wearing partial PPE approached us; one of them took the reins immediately and instructed one of the national health care workers to admit the baby.

In a few hours the baby, having tested positive for malaria, was being treated properly and sleeping. Days later, when I checked on her during this Saturday visit, I saw her lying in bed playing with her foot. Soon her mom would take her home. Without the triaging, she might well have never been seen or taken care of as an indirect effect of the Ebola outbreak's curtailing the treatment of common medical conditions like pneumonia, malaria, and tuberculosis.

Looking over the triage book at St. John's, I could see that very few suspected cases were sent to the ETC during my weeks in Sierra Leone, and that the ones that were sent usually turned out to be negative. We ourselves did send a few cases that fit the protocol for suspected Ebola on to the ETC, though at least one of them was a weak case. Even when the diagnosis was acute, our ambulance sometimes took more than an hour to arrive.

Patient traffic in the triaging rooms varied widely. One early afternoon a small flood of patients were suddenly triaged and taken care of, with only a trickle of patients walking in later. One of these, brought in by taxi, had been discharged from the ETC two weeks earlier after two negative tests for Ebola. He had felt fine the first week after discharge but then felt poorly, though the only symptoms we could get from him were weakness and poor appetite. He did not look dehydrated to me. We decided to send him into the outpatient clinic. While he was being triaged, he was given an energy bar, which he devoured eagerly by squeezing the bar out of its wrapper with his teeth. Our conclusion: he did have an appetite—he was just weak with hunger.

BY THE FIRST week of March 2015, Liberia had made real progress in the fight against Ebola, discharging the nation's last Ebola patient and reporting no new cases in the preceding week. Sierra Leone and Guinea, on the other hand, reported a combined total of 137 new cases, an increase of thirty-four compared to the previous week. The virus was still widespread in Sierra Leone, with eighty-one new cases, including twenty-six in Freetown.

Back at the ETC, during the first few days of March, we saw a substantial increase in the number of new patients, the census standing at fifteen on March 3, eleven on March 5, and nearly twenty on March 8. My buddy Dr. Kamara, who had been actively fighting this Ebola outbreak since the fall, commented on how sad it was to walk among the confirmed ward patients on the mattresses and know that probably only one out of the four would survive. The chances for infected babies and children under five were even slimmer. Although no one seemed to know the rate of transmission of Ebola from a mother—in utero or at delivery—we definitely saw children from such mothers who tested negative. We operated on the idea that transmission was not absolutely certain. Fortunately a large proportion of our patients never got beyond the suspected ward, and most of these quickly tested negative for Ebola. They and the probables were all in stable condition and usually sat outside after their breakfast so that Sulaimen and

I could accomplish the majority of our rounds in the open air, over the fence. While we were doing this, I kept thinking how much such interviews contravened privacy constraints in the USA, such as the HIPAA provisions.

After March 8 the census fell substantially and remained low during my remaining days in Sierra Leone. No new Ebola cases were confirmed at the Lunsar ETC after March 5, and, unsurprisingly, many of the suspected cases turned out to have been malaria. That hardly meant we had the outbreak under control. On March 13 a fourteen-weeks-pregnant woman was picked up from a quarantined home, bleeding from her womb. While sitting on a stone slab in her village, she had fallen backward and hit her head. She expelled her fetus during her first night in the probable ward, though her placenta remained, and she died not long after. Her Ebola test came back positive the next day, which meant a lot more people in the quarantined home had probably been exposed.

Another pregnant woman in the wards, RK, had better luck. Ten weeks pregnant, she had been there for thirteen days and had recently shown herself as relatively asymptomatic; in fact she had been so famished that she asked for more food. One night as we did rounds, she was sleeping soundly when we let drop a mention of food. Immediately she jumped up as if looking for it. It took her a long time to clear the virus, and she was unable to leave the ETC for another five days, when she was transferred to a Doctors Without Borders facility in Freetown for the rather sad but unavoidable termination of her pregnancy.

As always, children in the ETC moved me to tears. One day we admitted a seven-month-old baby who had symptoms; the mother, who was not a patient, decided that she would stay with her and run the risk of exposure. A similar situation arose with Marie, a four-year-old girl sent to the ETC by triaging at St. John and admitted into the suspected ward. She was accompanied by her mom and her aunt. The mother, who had another baby at home still nursing, was none too happy to have been sent here and refused to have anything done to her daughter, including drawing blood. After a lot of persuading and explaining, we admitted the girl; her aunt insisted on staying with her in the wards.

The next day Marie was the one who looked unhappy as she sat on a chair looking at her mother in the visitors' tent across from the orange fence. The separation might as well have been a vast ocean. Marie was not very ill, but when the national nurses came to give her medicines, they had to more or less force the pills into her tiny mouth, holding her arms down while she was lying in a supine position. I was afraid she might aspirate, but in the end she threw them all up. I was appalled at the way Marie was treated, but the nurses nonchalantly brushed it off, saying that was how children in Sierra Leone were always given medicines.

ON ONE FRIDAY that I had off, I hitched a ride to Freetown. The capital was hot and crowded, and I recognized some of the landmarks from being there four years earlier with my son Charles. Legend has it that the cotton tree in the center of town is the tree where in 1792 African-American former slaves gave thanks after landing. Not too far from there, the black settlers built St. John's Maroon Church in 1808, using beams from a slave ship for the ceiling.

It being Friday, hundreds of men and some women had spread their prayer rugs and were saying prayers in front of the big market. The market sold baskets, voodoo dolls, trinkets, handbags, fabric, animal skins, African masks, and some quite menacing-looking carvings. The two-story building was dimly lit, and some of the sale items seemed to be gathering dust. I saw no other customers, perhaps because it was lunch hour and the hottest part of the day.

The Old Wharf, or "Portuguese Steps," had been washed away about two years ago. Hamid, the driver, took me down a very narrow set of broken stone stairs flanked by stone walls and dilapidated huts where partially clothed children were playing. He carried my backpack protectively, reminding me that there were many "bad people" here and I had to be careful with my belongings. At the end of the winding stairs, we walked through someone's kitchen, and to our left was a dark passageway; then, right when the steps ended abruptly, we had a magnificent vista of the Atlantic Ocean. Sea breezes blew briskly; those huts lining the stone stairs were surprisingly cool, I thought. Whoever lived there had their own beachfront property!

We passed through the slums of Freetown with tin-roofed hovels that stretched far from the road to the edge of the ocean. Massive garbage dumps were piled up alongside a row of houses, smoldering in a slow burn. Many of the canals leading to the ocean were choked with garbage, and pigs rummaged happily for food. It was a wonder that Ebola would leave untouched some communities of people living in such close quarters.

I WAS BEGINNING to get known, even in town. One morning at the market in Lunsar center, as I was surveying the produce and dry goods, someone called me by name. When I turned, I saw a familiar face who asked if I was shopping. After exchanging a few words, she then inquired, "Do you remember my name?" At the treatment center, "What is my name?" had been an ongoing game that the nationals liked to play with me. It was difficult for me to remember both faces and names, but I worked hard at it. I told the woman in the market that her name was Lucy, and she broke into one of the happiest and widest grins I've ever seen. Then she helped me to buy some cucumbers.

Abu Bakar came to pick me up. He was the driver who had been instrumental in teaching me some Temne, including such phrases as *"Tope ander-a?"* ("How are you?"), *"Seke"* ("Hello"), and *"Thanda kuru!"* ("praise God!"). He noted that I interacted with people at the market and said he felt that many expats were tight-lipped and kept very much to themselves. He asked what I would miss most after I left Sierra Leone. As I answered, "The people," a man walked right into the path of the cruiser, barely looking where he was going. Abu Bakar laughed and said life in Sierra Leone was hard and people were poor and hungry, always searching for food and jobs, and sometimes they did not pay attention on the roads. Ebola did not make it any easier.

He blamed Ebola for his wife's death the month before. She had begun to bleed after giving birth to their third child, but no one would help her for fear of getting Ebola, even though she had not been infected. (As in Liberia, many midwives had died from exposure to pregnant women infected with Ebola; as a result pregnant women in general were often seen as likely carriers.) Abu Bakar said she had fallen and died; as he held her in his arms, he wondered whether this would have happened if there had been no Ebola outbreak. By the time he finished telling me his story, we had reached the ETC, and he was so grief stricken that he could hardly contain himself. He covered his face with his hands. I knew we were to avoid skin contact, but I could not help but reach over and touch his shoulder gently. Another vehicle with a fellow driver came along from the opposite direction. Abu Bakar composed himself and acknowledged his colleague by nodding.

On one of my days off after a night shift, when we failed to find a vehicle early enough to go to the beach, a colleague and I instead drove to Makeni, an hour away in the Bombali district. There we paid a visit to the hundred-bed Mateneh ETC built by the British military, the treatment center that just a week earlier had experienced a resurgence of cases. Many Lunsar ETC staff had moved here, and it was nice to be welcomed by familiar faces. We missed the shade provided by the tin roof over our staff walkways in Lunsar, but the friendly person who took us around was evidently proud of the facility. Many of the confirmed patients who were on the road to recovery sat under covered walkways and waved. On the day of our visit, they had a total of thirty-four patients: fifteen in confirmed, eleven in convalescent, five in suspected, and three in probable. (Although

Lunsar also had a formal convalescent ward—a unit to house patients waiting for a second negative test—it was used only once during my entire stay there.) Because of the bigger patient load, the medical staff split up their duties. The mortality rate of their patients was about 50 percent.

Some portraits of other Lunsar patients:

ROSEMARIE & HOUSEMATES

Seventy-year-old Rosemarie had come to believe that we were attempting to poison her, so she tried everything she could to not take her medications. When given a bunch of pills to swallow, she would hold them in her mouth, gingerly swallowing water without swallowing the pills. As soon as we turned to leave, she would spit them out. The nurses had become suspicious after they noticed a bunch of pills lying on the cement floor. Soon Rosemarie was responding only to verbal stimuli and subsisting on IV fluids alone. As we could no longer make her take pills orally, she had got her wish. I wondered whether she truly believed we were responsible for her demise.

A day later Rosemarie had passed away, and before the week was out two women who had lived in Rosemarie's house were admitted and tested positive: Mbalu and her sister-in-law Marai. The two looked well enough when they arrived, but I was afraid it might be the calm before the storm of Ebola. Marai was tall and attractive, with prominent cheekbones, whereas Mbalu appeared unselfconsciously topless, breasts dangling the whole time she talked to us—a fairly long period, since we had to wait for a full liter of fluid to drip into her. This display must have been the natural thing here, since when she asked for a wrap it was to cover her head, and neither of the national health care workers doing rounds with me seemed even slightly perturbed. Physically Mbalu looked much like Fatmata K.'s mother—so much so that I had the strange premonition that she would end up suffering the same struggling and bloody death. Sadly my premonition proved accurate, and she died a few days later. Marai, on the other hand, remained without symptoms, and recovered.

YAEMA AND HER CHILDREN

The sad story of Deborah and her family had a tragic echo when another group of relatives of that family turned up at our doors. The first to arrive were two young brothers, Ishmael and Emmanuel, ages four and two, admitted from the observation interim care center (the holding center for quarantined persons) when their temperatures spiked after a two-day stay there. Their father had paid a visit to Deborah's family during the early stages of their illness; before long he too had contracted Ebola and died—but not before infecting his own family. Now at the end of February, two of his sons took places of their own in the Lunsar ETC—on the very day that Deborah, whom we had presumed to be the last of the extended family, passed away. Emmanuel was eating until he caught sight of us in PPE and stopped, visibly pulling away. He was feverish, restless, and resisted being fussed over, unlike his older brother, who ate heartily.

Two days later, two more members of the immediate family came in: the twenty-two-year-old mother
of Ishmael and Emmanuel, Yaema, and her five-month-
old baby girl, Mabinty. They had both tested positive
and joined the rest of the family in the confirmed ward.
Though at first the new arrivals seemed okay except
for some restlessness and nausea on Yaema's part, by
morning they both began to have bloody diarrhea, a
sign that did not bode well. Also Mabinty (as well as
Emmanuel) had developed a morbilliform rash, which
could have been caused by Ebola or measles.

Sure enough, Yaema lost her five-month-old baby,
Mabinty, within days. I was surprised to discern no trace
of sadness on her face, though I wanted to believe that
she was just fatigued and having to remain strong for her
two remaining children, that inside she was struggling
with her loss. Indeed, both Ishmael and Emmanuel were
weak, receiving IV fluids and being fed protein-rich
food through nasogastric tubes, as they refused to eat.
Their clinical status was touch and go. I discovered that
their great-grandfather, Emmanuel Senior, was also in
the confirmed ward. He looked elderly but still strong.

When the two-year-old Emmanuel died two days
later, Yaema again did not look as devastated as I imagined she ought to be. In fact it struck me that she looked
so young she might have passed for a teenager, and that
sometimes she behaved like one—sitting up late at night
to watch a movie, rejecting the food she was served and
asking instead for *fufu* and peppered soup. Her four-
year-old, Ishmael, pulled out his nasogastric tube a few
times; he was still struggling with diarrhea, but thankfully it was no longer bloody.

Meanwhile, Yaema's grandfather Emmanuel Senior was in the final hours of his struggle. He had difficulty with his breathing, and after helping him to bring up a thick mucus plug, we raised the head of his bed. He seemed more comfortable. I wanted to give him a low dose of morphine as well, but when I mentioned it, the nurse I was with looked mortified. I explained that it would further improve the old man's breathing. She remained unconvinced, instead asking me to help her lift the patient farther up the already inclined hospital bed. I complied after finding a big vinyl block to use as a pillow. The nurse then observed that he was breathing a bit better, which was true. She also declared that she did not administer morphine. The other national nurses who were in the confirmed ward at the time concurred, and I let the issue drop. Later I learned that many nationals, seeing patients who were given morphine usually die within a few days, concluded that doing so was somehow contributing to rather than mitigating the process of dying. Surely the morphine might have eased the old man's last hours (he stopped breathing altogether the next morning) but that would have only confirmed the nurses' superstition about the drug.

Shortly after Yaema lost her grandfather, I asked her how she was coping. This time I finally elicited some emotion from her. She told me that when her baby died five days before she had been quite ill herself and felt "mixed and confused feelings inside," that when her two-year-old died two days earlier, she had cried inside her heart. Only recently she had lost her husband as well. Besides that loss, she was now left with only two of her four children—a seven-year-old, I learned, was living with a relative in Freetown. So much tragedy in such a short period, and at such a young age!

Ishmael's condition fluctuated regularly in the days that followed. His diarrhea was persistent and dictated continuing his IV fluid, especially since he was still refusing to eat and pulling out his nasogastric tube. Whatever her sadness, his mother recovered quickly and was content to have the caretaker look after her son. We encouraged her to interact with him more, thinking he would be more interested in eating if his mother were involved. In fact he did improve, although he groaned and whined a great deal. We were gratified to see him lying on top of his mother's belly outside the ward, shaking his head when we asked him whether he had such and such a symptom. We went through a whole list of food, trying to find something enticing that he would be willing to eat, but the only one he nodded to was peppered soup. The kitchen did not prepare his peppered soup the next day, and again he refused to eat, even though his fever had disappeared.

When Ishmael had his first negative Ebola test a few days later, everyone cheered. They cheered even more loudly when he and his mother finally left the ETC after we got news of a second negative test. The three long weeks he had been there must have felt like an eternity. For us, his departure felt like the victory everyone had been waiting for: we had had few Ebola-free discharges, especially not discharges for someone as young as four. I was at St. John and could not share in the celebration but was glad to have been a participant in his care.

Alas, the discharge did not mean that Ishmael was out of the woods. His immune system had been ravaged by Ebola, and he was still weak. A day after the discharge, we learned that Ishmael, according to a secondhand source, was "trying to die," that his mother had refused to take him to Port Loko Hospital and he had been brought back to the ETC. He died the night before my final day in Sierra Leone. Yaema was down to one surviving child, the one in Freetown. Not having contact with his own family had saved him from near-certain death.

VII

EBOLA AND US

WHILE WORKING IN West Africa, I often reflected on the gulf between American and African attitudes to the Ebola crisis. In the fall of 2014, while the epidemic was rampant in Sierra Leone, many of its own citizens did not believe that Ebola was real and that it was also often fatal. Yet when a single infected person reached the shores of America and later spread the infection to two nurses at the hospital that had first misdiagnosed him, panic spread like wildfire. Drastic measures were quickly imposed—including the quarantine for twenty-one days of returning volunteers who displayed no symptoms. Commentators called for a ban on incoming flights from West Africa, including Donald Trump who famously tweeted way before Ebola reached America, "The U.S. must immediately stop all flights from EBOLA infected countries or the plague will start and spread inside our 'borders.' Act fast!" I even read one proposal that returning volunteers be detained in a vessel off-shore!

By New Year's Day 2015, some progress had been

made in containing the outbreak within the three affected countries, although there were still ups and downs. So it hit home hard when in the span of a few days in mid-March we in Lunsar learned that three volunteers in Sierra Leone—one American, one British, and one a Sierra Leonean national—had contracted Ebola. The American volunteer worked with Partners in Health (PIH), the humanitarian organization cofounded by Paul Farmer and based in Boston. The volunteer was first transferred to the Kerry Town ETC for health care workers, then flown back to Bethesda for treatment. The national health care worker also worked with PIH, in the same ETC as the American, in Maforki chiefdom in Port Loko district. He was admitted here to our Lunsar center. When I told our medical coordinator that I was puzzled as to why the national health care worker had not been admitted to the treatment center in Kerry Town meant specifically for health care workers, she said something to the effect that one had to apply for admission there. Other of my colleagues asked the same question and got the same answer: admission to Kerry Town was not automatic; an application had to be made and approved. By whom was not made clear: The British military or the Sierra Leonean government?

Naturally, we found this new patient's arrival disturbing. We had regularly brushed shoulders with many of the health care workers from PIH at our weekly roundtable conferences. We were in this fight together, and when one of us was down, we all felt it. Another potential concern was that after a long period of calm with no Anglophone health care worker contracting

Ebola, a renewed media frenzy around the infection of the American or British volunteer might once again throw the general public out of kilter. My brother sent me an email wondering if I was the "US care worker down with Ebola" and saying he wished I were already safe at home in the USA.

Our first concern, however, was seeing that Usman, the national health care worker who had contracted Ebola, recover as quickly as possible. Watching him get out of the ambulance and walk the lonely path into the confirmed ward was deeply unsettling to all of us. We, fellow health care workers, greeted him wholeheartedly and encouraged him to think on the positive side. In his mind he must have rerun every possible scenario, trying to pinpoint when the critical breach had taken place. To have witnessed all that Ebola patients had been through—the horrendous outcomes for so many, the intermittent triumphs—and now to be on the other side of that fence: how surreal and scary that must be! We spent a good deal of time reassuring Usman that we would do our utmost, but that he would need to have faith. I reached out to give him a parting pat. Looking into his red eyes I saw fear, resignation, reassurance, gratitude, and a sense of unity. For a brief moment as I walked from the backyard into the ward, my eyes moistened from the picture I had of myself in his shoes—would I fare any better? He was experiencing the nightmare of every Ebola fighter. To date, 840 health workers in West Africa had tested positive for Ebola; 491 of these had died. In the days immediately after Usman's arrival, all of us seemed to pay a little more attention to how we conducted ourselves. Our steps became more measured and deliberate, especially during the doffing process, probably the most likely time for breaches to occur.

A group of us doctors, including one from Sierra Leone, remained concerned that a Sierra Leonean national had risked his life working alongside the American, yet when both came down with Ebola, the American was sent to Kerry Town and then medevaced to NIH, whereas Usman was routed not through Kerry Town but to a smaller, regional center where the most aggressive possible treatment was vigorous IV hydration. As the disease progressed, Usman began to have frequent diarrhea, with fairly significant abdominal pain that required morphine. We gave him plenty of ORS and IV fluids, but I could easily picture how he might weaken significantly in the next several days.

Was a national's life considered less important than an American's? Soon, American media reported that ten more Americans from the same PIH Ebola facility as our national had also been evacuated to the US. Although currently asymptomatic, they apparently might have been exposed, and it was felt desirable that they be close to facilities that could administer advanced treatments immediately if the need arose. We found this news hard to harmonize with the situation of Usman, who was both infected and symptomatic—and in great danger—yet remained in our humble ETC. Usman often fingered with a great deal of tenderness a sheet of paper with the picture of an American woman volunteer who had sent her prayers to him on behalf of PIH. Might he not feel abandoned by his colleagues? Expats working in any humanitarian organization often seemed to enjoy better working and living conditions than nationals. Were they second-class citizens in their own country? On the PIH website, Paul Farmer, the cofounder of the charity, is quoted saying, "The idea that some lives matter less is the root of all that's wrong with the world." How did one reconcile this noble ideal with the reality of this particular patient's situation?

While these questions festered among some of us, I was the fortunate recipient of some complimentary press coverage. A few weeks earlier, Jenni Marsh from Hong Kong's *South China Morning Post* had interviewed me. That interview became the cover story of the March 15 Sunday magazine: "Behind the Mask: One Doctor's Experience on the Ebola Front Line." For me it was satisfying to see the story reach Asia, where I was born and grew up.

Two days after Usman's arrival, we got word that he was finally to be transferred to the Kerry Town ETC. Why it took two days to get him admission there was a mystery, and at first Usman actually refused to go, saying that since he was at Lunsar already, he would just as soon stay. Later he came around, all the while praising the good experience he was having in our ETC. In the late afternoon, as we were doing rounds in the confirmed ward, we heard that the ambulance had arrived to take him to Kerry Town. Again Usman hesitated, and a national doctor had to talk him into the transfer by emphasizing that Kerry Town had the ability to monitor his electrolytes. Very reluctantly, he agreed. Before beginning the transfer, six of us gathered around him decked out in our suits of PPE— all nationals except me—and urged him to rally his strength, which he would need in the coming days. I will always remember this tableau of solidarity for one of our own. While the others then got busy getting things ready, I spent a few private moments holding Usman's hand and offering the encouragement I would have wanted in a similar situation. The IV fluid was left hanging; just before departure we gave him a shot of morphine for his abdominal pain.

The next day I was on night shift, and during the day a group of us traveled to Bureh Beach. On our way to the beach, we passed the Kerry Town treatment center and wondered briefly whether we should pay our sick compatriot a visit, but decided against it.

On the beach were some Danish volunteers from GOAL, the Irish NGO devoted to helping the poor that was also assisting in the Ebola battle. They were enjoying their last day in Sierra Leone after completing their voluntary assignment. A couple of them were quite upset about the American doctor who had contracted Ebola. Apparently GOAL and PIH volunteers lived in the same compound in Port Loko, and according to them, the American doctor had not been careful in following procedures, and when he fell ill several people—including three of their Danish colleagues— had been exposed and required to be evacuated back to Denmark. Not knowing that I too was an American, they told me that all the Americans had gone home and the Maforki Ebola treatment center now had only four patients and was being run only by the nationals. I was sure it must have been disappointing to the evacuees to have to leave without making a contribution, but even so, I thought my interlocutors should have been more forgiving toward the American volunteer, who was then listed in critical condition.

We drove home in the midafternoon to get back in time for our night shift. I fired up my computer and noticed an email from Sheri Fink, a *New York Times* reporter who had spent some time at the Bong ETU while I was in Liberia. She was writing a story about the infected national and wanted to ask me questions about him. She had learned that his name was Usman and he had worked at Maforki along with the infected American health care worker. I told her about my surprise that he had not been sent immediately to the Kerry Town center like the American. I also told her some of my thoughts about the difference in treatment between foreign and national health care workers, and that perhaps we should ask ourselves to look more honestly at the moral dilemma posed by this issue.

The next day the *New York Times* published Sheri Fink's article under the title "Care Differs for American and African With Ebola." The day after that, all hell broke loose. I had just finished my night shift, the heat in my room was slowly building, and I was trying to get a fitful rest. My cell phone rang: it was the medical office calling with the message that I was to meet at once with the medical and nursing coordinators at the restaurant in Bai-Suba about an urgent matter. My heart sank, and I felt as if someone had punched me in the stomach. I had not yet read Sheri Fink's article, but deep inside I knew what they wanted to talk about.

The air-conditioning was off at the restaurant, too. With grim faces, both coordinators confronted me with the fact that I had been told not to talk with the media without first obtaining permission. In short: I should never have communicated with a reporter, much less a reporter from the *New York Times*. The nursing coordinator sat sideways and avoided looking me in the eyes. The medical coordinator then told me that it was imperative that I shut down my blog before the afternoon was out.

"On whose order?" I asked. She paused, then glanced at the nursing coordinator, who muttered unconvincingly that the order had come from the directors of both the regional and national offices of the IMC. Of course I had no idea who those people were, but I have a powerful rebellious streak as well as a strong sense of right and wrong. A feeling of injustice welled up inside me. I invoked my right to freedom of speech and declared that before I took down my blog, I would need to hear directly from the directors themselves. Furthermore, weeks before I had given the US media director the link to my blog and I heard no objections from her; I wanted to talk to her as well.

My blog seemed to particularly irritate the coordinators. "What are we going to say when patients go to your blog and read about themselves?" they asked.

The names had mostly been changed, I told them.

They were still identifiable, they claimed, and declared, "Ebola is quite a stigma!"

This whole idea was far-fetched. The truth was, ordinary people in Sierra Leone were struggling to survive, and they did not have ready access to the internet.

The medical coordinator looked taken aback. Apparently she thought I would meekly follow her order. She walked out of the room to make a phone call. I could not tell whom she was talking to, but after a while she came back and informed me that I would hear from someone that afternoon with further instructions.

Back in my room, I had trouble getting any rest and dreaded the impending call. When the phone rang late that afternoon, it was the head of mission, who wanted me to pack my bag and come to Freetown that very evening, even though IMC policy forbade travel after four thirty P.M. I asked whether he meant I should pack everything and leave Sierra Leone altogether, but he would not say. In fact I was so weary by then that I could not envision myself being organized enough to pack all my stuff. So I told him I would plan to stay overnight in Freetown and return to Lunsar the next day.

I arrived in Freetown after a three-hour trip in rush-hour traffic. I was met by a person from human resources who had arrived to take up her two-month position one week earlier. The next day, in the head of mission's office, we Skyped with IMC's media officer in the US, who alleged that the *New York Times* article had offended several of their partners: the Sierra Leone government, the British government, and PIH. Now they had to begin the delicate task of repairing relationships. To be quoted in the *Times* ensured my comments a worldwide readership, they said. The head of mission said that the night before he had been called in by one of the ministers, possibly the minister of communications, and asked to explain the *New York Times* article. It was hard for me to believe that a lone volunteer voicing her opinion could easily wreak so much damage to the national reputation. A protocol had to be followed, I was told, before the national health care worker could be admitted to Kerry Town. To hell with the protocol, I thought, we were talking about a valued health worker's life! The head of mission also noted that even Doctors Without Borders, which was generally known for its outspokenness, in matters involving a host country, had recently held off criticizing the government so that they could work better with its officials.

To my relief it turned out to have been the new person from human resources and not a higher-up who had asked me to shut down my blog. I found it sad that my coordinators had resorted to lying in their determination to muzzle me, but I was glad to have challenged them. In any case, I would have packed my bags and left rather than shut down the blog. As I waited for my ride back to Lunsar, I texted the medical coordinator to tell her I would be back at the ETC in the afternoon. This news must have come as a shock to her, as one of the volunteer doctors had called to tell me that she had announced at breakfast that I would not be returning.

Unable to rid themselves of my blog, the Lunsar coordinators made sure it would no longer have anything to say about the Lunsar ETC. That evening the medical coordinator came to my room to deliver my new schedule which exiled me to the screening and referral unit at St. John for the remaining week of my time in Sierra Leone. This effectively prevented me from blogging about the patients in the ETC. To quote one of the American volunteer doctors, "They have taken our most experienced and knowledgeable doctor and only infectious disease specialist and banished [her] to St. Johns." The coordinators who demanded shutting down my blog were promoting an Ebola information blackout. I told my medical coordinator point-blank that I thought it unconscionable for us to suppress the stories of the Ebola sufferers and survivors.

The entire imbroglio about the *New York Times* convinced me more than ever that the world will only get better if we are honest with ourselves and confront the truth bravely, however uncomfortable or unpleasant it may be. The fact was that the infected national health care worker had as much right to be admitted immediately to Kerry Town as the American and British health care workers, but he had been denied immediate admission. It was an inconvenient truth that the NGOs needed to confront and address. When the NPR story about the Bong ETU in Liberia was broadcast—based on my blog as well as interviews with me—IMC never asked whether I had obtained permission to talk to the NPR reporter. Not only had they had no issue with it, but they posted a transcript and audio of the story on the IMC webpage, including the names of patients in print. Apparently media coverage was acceptable only when it painted the NGO in a good light.

On his website, Partners in Health cofounder Paul Farmer only briefly mentioned "the infected national" in context with an announcement that an additional four PIH volunteers who had had contact with him were being sent home to be monitored, five days after the infected American was evacuated. Beggars cannot be choosers, but in the world of the humanitarian organizations, the beggar—that is, the government of the country that needs aid—usually dictates the terms of how and where the NGO can work. Either the

NGO abides by these terms or is not allowed in. Few humanitarian organizations dare openly criticize the government of the country in which they hope to work; the only notable exception may be Doctors Without Borders.

VIII

TOWARD
ZERO EBOLA

I SPENT MY last days in Sierra Leone aware that some of the IMC administrators wanted very much for me to go away. I knew for sure that my blog posts were being closely scrutinized, so I titled the first of my post-"scandal" entries "The Reports of My Death Have Been Greatly Exaggerated." Those in the know would be sure to understand—and I was quite sure that, despite their supposed objections, they were probably following my blog faithfully every day.

In a matter of days, I would be leaving Lunsar for home, sweet home. On one of the last evenings, the rain clouds gathered and the wind picked up, and for a moment we hoped we might get a shower to break the dry spell, but the clouds soon dissipated as quickly as they had gathered. The next morning a pitter-patter was heard on the roof of St. John, alerting us to a passing shower that brought no relief and only added a dose of humidity.

Esther and Dorcas, the two Nigerian nurses at St. John's, were representatives of the African Union. They worked as a team triaging patients in Krio, the creole language derived partly from English. By this time in my stay, I had reached a point where I could understand some of this language. Esther in particular was quite shrewd in questioning the patients about possible exposure to Ebola and did so with great humor, hypothesizing scenarios suggested by the patients' occupations.

Since my last visit, the pediatric ward had filled up, a good sign. The villagers had regained confidence that the hospital had taken adequate steps to prevent transmission of Ebola infection to bring their children for care. Some cases involving children and newborns were heartbreaking. I remember one six-month-old baby brought in by his fatigued mother wrapped up in a worn *lappa*. Instead of regularly breathing, every so often the baby would emit a throaty, rattling breath. He looked lifeless. His mother said the baby had had trouble breathing for two days, but according to her had still been nursing well as recently as this morning. It was hard to believe a barely conscious child could be nursing, and for a while we were hesitant about what to do. We considered letting the mother and baby sit in a corner until he drew his last breath, a more peaceful end to what looked like a losing battle. However, a doctor from the hospital seemed to think we should admit the baby. The inpatient department tried to resuscitate him for at least three quarters of an hour with an Ambu bag and CPR. His pulse returned, but there was no ventilator to keep him going, and it was not feasible for someone to continue manual resuscitation for a prolonged period of time. The nurse placed an IV and a nasogastric tube. Eventually there was no spontaneous respiration, and the resuscitation effort stopped. The baby fell back into his earlier irregular rattling breath, and there was still a flutter of a heartbeat. The grieving mother was told there was no hope. She was tearful and sat quietly and alone in a corner with no family and no baby (fathers were rarely seen in these situations). Once the baby's breathing and heartbeat stopped, the hospital would inform the burial team; the mother would not be allowed to touch the child before interment.

Few of the many children and pregnant women we saw presented likely cases of Ebola, although two sick babies who came in together with their moms fulfilled the symptom criteria and were candidates for the ETC. However, they had had no contact with infected individuals, and in ordinary times they could just as well have been any baby with gastroenteritis or malaria. But these were not ordinary times. The two mothers were both robustly built and vocal. After some time waiting in the corner together, they came to the conclusion that they would not let their babies go to the ETC, even though our triage team argued that they should be transferred there for testing. I too would probably have protested if I had been the mother. A few days waiting for test results at an Ebola treatment center always brought the risk of exposing a child to the infected. Besides that, the babies would be separated from their moms for days without any physical contact. In the end, a government surveillance team came and agreed to let them be monitored for symptoms at home. The babies were discharged with antipyretics and antimalarial medications. A couple days later, one of the babies was in the hospital being successfully treated for malaria.

The case that preoccupied me the most during the last days at St. John's involved a twenty-five-year-old mother and her emaciated one-month-old baby. The baby arrived all wrapped up in a *lappa* and wearing a colorful winter hat. But after the mom slowly unfolded the *lappa* to reveal the baby, we stood there stunned, speechless: the infant was starving—her face was aged and wrinkled, skeletal arms and legs and skin stretched tautly over a prominent and delicate rib cage. We found out that this was the woman's second child. Though the mother did not look particularly malnourished, she said she had not been producing sufficient milk. The crying baby was eager to suckle when put to her mother's breast. We sent both to the inpatient unit to be admitted.

In the afternoon I went to the pediatric ward in search of the infant. I found the tiny baby with two nurses hovering over her, skillfully placing an IV into her thin arm. As a nurse took the baby to be weighed, she cried loudly and we saw clearly for the first time that the baby had a cleft palate, which might well have complicated her attempts to breastfeed and contributed to the malnutrition. The baby weighed only four pounds.

A nasogastric tube was provided for feeding, and the mom was handed a cup and asked to express some breast milk. I asked the nurses whether they had any infant formula available in case the mom was not able to produce enough milk. The hospital had no milk, they told me, and no one knew when or if there would be a shipment. Perhaps one of the feeding centers in Makeni or Port Loko was open, but that would entail discharging the baby from St. John and asking the mother to find her way there by public transportation. The logistical nightmare that would impose on the poor mother might well kill the child.

So I decided to call the ETC and ask if they had infant formula to spare, explaining the reason for my request. The medical coordinator was frosty: she was not sure whether or not they had any formula and even if they did, she did not think "the donor" would approve its use for purposes other than Ebola treatment. Why not have the baby transferred to Port Loko Hospital? she suggested. I told her that the staff at St. John's was not sure if the feeding centers at Port Loko or Makeni were open, and that I was sure that the donor would forgive us for redirecting a tiny amount of the formula to save the life of a starving infant. I emphasized that the medical director of St. John's had confirmed that the hospital had no formula, nor a schedule for getting any. She would talk to him herself, she told me, then she would check with the medical director of the ETC to see if they could spare me a small amount. I told her that if this was not possible, I would go to Lunsar pharmacy myself to buy the formula. In the end the ETC spared me between a half and a full day's supply of infant formula, which I delivered in the late afternoon, while the nurse patiently gave the baby a syringeful of expressed breast milk through the nasogastric tube.

The next day the starving baby was alert and drinking from a cup held by her mother. The mother had been able to express breast milk for her, so two-thirds of the infant formula was still left. The infant kept looking alert and suckling eagerly for two more days, although she remained very pale and her hemoglobin was only 6.8 grams per deciliter, so she also received a blood transfusion.

On my last day in Lunsar, I went to pay a visit to the starving infant and her mother, only to learn from one of the nurses the sad news that the infant had died the evening before, most likely from malnourishment. Her mother sat on the bench quietly, staring blankly and despairingly into space. The ward where the baby had been was now emptied of patients. Through the screened window I could see the baby swaddled in a *lappa*, lying all alone in the big bed. Death among the little ones seemed so commonplace here. Surviving infancy meant dodging a gauntlet of circumstances thrown up by poverty and disease. A day earlier the baby had been suckling with some relish, and then her short life suddenly ended. Her mother was not permitted to hold or hug her a last time: the rule of Ebola.

MARCH 23, 2015, marked one year since the WHO had confirmed that the hemorrhagic fever killing people in remote parts of Guinea was caused by the Ebola Zaire virus. Since then, Ebola had infected 24,701 people and killed 10,194 in West Africa. April 15, 2015, was the original target date for the three Ebola-affected countries to reach zero new cases, but as April approached, it was clear this goal would prove impossible to achieve. International mobilization had broken the back of the epidemic, but sporadic outbreaks kept occurring, and it would probably take until summer for all three countries to claim realistically a near zero result. Liberia was typical: after going Ebola-free since the beginning of March, a forty-four-year-old woman was reported to be diagnosed with the infection. The cause was presumably sexual intercourse—her boyfriend was an Ebola survivor.

On March 21, the day after Sierra Leone recorded its first zero-number day in nearly ten months, President Ernest Bai Koroma announced the launch of a four-week "Zero Ebola" campaign: a national lockdown that called on all six million Sierra Leoneans to stay home for three days at the end of March and for three consecutive Saturdays in April. No markets or large gatherings would be allowed, though in March there would be a window for people to attend Friday prayer at a mosque or church on Palm Sunday. During the lockdown, surveillance teams would look for symptomatic individuals. A surge of patients was expected at the Lunsar ETC.

ON MY LAST day off, I finally got a chance to visit River No. 2, a beach right outside Freetown run by the community where all income earned was plowed back into local improvement projects. The only reason the head driver was amenable to getting us a vehicle early enough to make it to the beach was because one of our trio was a Sierra Leonean; he even saw to it that the cruiser was fueled up. The other colleague who came to the beach was a nurse from Kenya who had been in Lunsar for close to two months but had yet to make it to the ocean. Beach outings were commonly organized by an expat, and if you were not in her or his group, you were not likely to be invited to join.

The driver drove first to Freetown and then the beach, and when we got there, the tide was going out, creating a river that physically separated our beach from the adjoining one. The river currents were too swift to swim across, so two wooden boats ferried people across at frequent intervals. Unlike Bureh, this beach was frequented by many expats, including families with young children who came with their nannies. The waves at this beach were gentler and the temperature of the water perfect. Many vendors plied their wares on the beaches: dresses, skirts, bags made with African fabric, carvings, trinkets, and lappas. There were also a fair number of provocatively and scantily clad young ladies to represent the world's oldest profession. No matter the trade in goods or flesh, we three agreed that the outing was a perfect release from the stress of the ETC.

Instead of going through Freetown on the way home, the driver took the coastal road along the peninsula. For an hour we traveled on bumpy red dirt and across three very narrow bridges with missing guardrails. Luckily we were spared a plunge into the water, but we did see the rusty remains of a huge truck lying on its side on the boulders of one creek bed. The Chinese had built new, smoother roads in Sierra Leone, but on the other side of the peninsula. Up high on one of the mountains along our route, the driver pointed out a somewhat weather-beaten mansion originally owned by the Chinese government. Years ago one of the nation's politicians had asked the Chinese to give it to him, and they did. He quickly tired of the gift, however, and after being used for a period by the military it had been left unoccupied and untended, and was rapidly deteriorating.

That evening I got a further dose of therapeutic calm from visiting the vegetable garden in Bai-Suba. Here

I enjoyed walking through the rows of Irish potatoes, beans, carrots, turnips, cabbages, eggplants, tomatoes, lettuce, onions, cucumbers, and watermelons. I worried that many of the new seedlings would shrivel up and die in the hot sun. With no irrigation system, all watering had to be done with watering cans, and by midmorning workers stopped for the day. For some strange reason, the sheep and goats never seemed to wander in to feast on the vegetables.

While I packed I found three big, empty plastic bottles. Trash, especially plastic, is a huge problem here, and I was reluctant to add to it. I had tried to reuse the bottles, replenishing from the water cooler in the ETC, but at St. John there was no water cooler. I walked out and found some laborers who were building drains under the unforgiving sun. I was a little embarrassed to give them such meager offerings as three empty bottles that needed refilling, but they were eagerly accepted and in the end I felt bad I could only make three of them happy.

I said goodbye to the WASH and triage teams at St. John, then went around the ETC saying my farewells to the medical, kitchen, and pharmacy staff. There was only one patient in the confirmed ward. No new cases had been confirmed in the past week; the patient census was also down in Makeni, and the Lunsar staff who had been sent there to help had all returned.

During the departure routine from Lungi Airport, all passengers had their temperatures taken at each stop—at the speedboat ticket office, airport entrance gate, and airport check-in. At a further health station in the airport, we filled out questionnaires about symptoms and where we'd been and what we had been doing in Sierra Leone. The gentleman ahead of me set off an alarm on the infrared thermometer with a temperature above 38 degrees centigrade (100.4 degrees Fahrenheit). The tester used three different thermometers on himself and then tested the man again, who repeatedly set off the alarm. The gentleman was pulled aside and taken somewhere else. That was a scary moment. I could not help but think, what if it had been me? We had arrived at the airport at least six hours early, and the man did not reappear until a couple of hours before departure.

First stop Brussels, where we filled out a public health locator form again as our temperatures were taken. The screening at Washington Dulles was much better organized than the year before. This time a group of us waited in a room to be called in one at a time. My first interview was with an officer in face shield, mask, and gloves who obtained my contact information, then put me in a room again—with stainless steel bench and table, but no sink or toilet. The officer kept the door open and politely offered me a proper chair in lieu of the cold metal bench. There I was quizzed by CDC officers, who had a lot more questions. Specifically they were interested in people coming from Port Loko, where the American health care worker had been infected. I did not make it clear to them that Lunsar was actually *in* the Port Loko district, only a half hour from the Maforki ETC.

After an hour's wait, I was given a packet of Ebola information and a sheet to record my symptoms and temperature for the twenty-one days of quarantine, a digital thermometer, and a cell phone with the CDC phone number in it. I retrieved my bag and sailed through customs and onward to the next leg of my journey home to Boston.

A few days later, I learned that the British health care worker who contracted Ebola at the Maforki Ebola treatment center, a military nurse, had recovered and been discharged from London's Royal Free Hospital after receiving an experimental drug, MIL 77. The American was still at NIH in serious condition. (He recovered not long after and was released.) The greatest news, however: Usman, the infected Sierra Leonean health care worker, had regained health without ever receiving any special drug and been discharged from Kerry Town. He was an Ebola survivor! *Thanda kuru!*

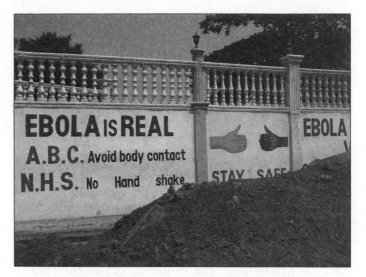

Ebola street signs in Freetown, Sierra Leone.

The medical team ready to go into the Ebola Treatment Center in Lunsar. Lai is second from the left.

Lai in personal protective equipment at the Ebola Treatment Center in Lunsar.

A slum in Freetown.

AFTERWORD

THE BIGGEST EBOLA outbreak in West Africa, spanning from 2013 to 2016, ultimately infected over twenty-eight thousand people, took the lives of eleven thousand, and spread to a handful of people in Europe and the United States. The epidemic peaked in October 2014, around the time I arrived at the ETU in Bong County, Liberia. The cases began to decline slowly after an upsurge of concerted international response. It was declared no longer an emergency on March 29, 2016, by the World Health Organization. At the end of March, WHO also declared the end of Ebola transmission in Sierra Leone, and in June 2016 declared the same for Guinea and Liberia, but it cautioned that these three countries "would still be at risk of Ebola flare-ups, largely due to virus persistence in some survivors, and must remain on high alert and ready to respond."

A smaller Ebola outbreak occurred in the Democratic Republic of Congo in May 2017, but this was quickly brought under control and declared over by WHO in July 2017, with four people dead and four survivors.

The West Africa Ebola outbreak left behind an estimated ten thousand–plus survivors who required care for post-Ebola syndrome. They have symptoms of joint and muscle pain, reported in 50 to 75 percent of survivors; eye problems, including blindness; hearing loss; and various neurological problems such as memory loss.

These constellations of symptoms can be so debilitating that survivors are unable to work.

Scientists have also discovered that the Ebola virus can persist in immune-protected sanctuaries such as the eyes and the testes. It has been found in semen and breast milk for over a year after infection. Sexual transmission has been reported. Infected men have to abstain from sex or practice safe sex until it is proven that their semen is free from the virus.

An experimental vaccine, rVSV-ZEBOV, was given in a few trials to contacts of Ebola patients and contacts of contacts. It appeared to be 70 to 100 percent effective in the prevention of transmission, but firm conclusions could not be reached because of the trial designs. The durability of the vaccine is between one to two years.

Overall, this Ebola outbreak showed that the international community lacks the capacity to meet a critical, severe, sustained, and geographically dispersed public health catastrophe. The WHO, governments, and many humanitarian partners were overwhelmed by unprecedented demands driven by ingrained cultural beliefs, poor or nonexistent health and local infrastructure, and other logistical challenges. The clash between a culture that promotes physical contact with the sick and deceased and a virus that infects on skin contact created a volatile situation that evaded conventional control measures and enabled the Ebola infection to spread unchecked.

Cultural beliefs will not change quickly, but this need not delay the funding necessary to bring health care facilities in poorer countries up to standard and to train more of their people to become health care workers. Preventive efforts as well as ongoing research must continue to find a vaccine cure for Ebola. Full accountability of designated disaster funds, including bringing corrupt officials to justice, is another prerequisite for any future international relief project. (In Sierra Leone a third of the funds for the fight against Ebola went missing.)

This seminal and historical event should serve to galvanize international cooperation. WHO certainly learned to respond more quickly to the Zika outbreak and worked with countries affected by it to make plans to coordinate efforts on finding rapid diagnostic tests, therapies, vaccines, and preventative measures. However, when the US congress failed for months to authorize emergency funding for Zika in 2016, the administration shuffled resources from Ebola- and other health-related projects to fund Zika research. The business of dealing with Ebola, however, remains unfinished. The three countries most affected by the Ebola outbreak need enormous help in building and reinforcing their health care infrastructure, and training more health care workers to improve the ratio of health care personnel to population. These are crucial to achieve a sustainable health system capacity, and if such funds are not restored, we might wind up repeating the horrific situation we recently overcame.

WHO has also responded expeditiously to the most recent Ebola outbreak in the Democratic Republic of Congo (DRC), reported in May 8, 2018; the ninth outbreak in the DRC. More than 7,500 doses of the experimental vaccine, the rVSV-ZEBOV Ebola vaccine, would be given via a ring vaccination technique involving identifying and offering vaccine to contacts and contacts of contacts of those likely to be infected in an effort to contain the spread of the virus.

For all these reasons, the stories of Ebola-affected victims should be told and remembered. As individuals we should not turn away from uncomfortable horrors, but summon the courage to heal with our hearts those we cannot touch with our hands. The Ebola outbreak took an immense human toll. Many who perished were buried in graveyards with temporary markers that will invariably erode in the harsh tropical climate, leaving no trace. We must remember both those who died and the frontline Ebola fighters who perished while caring for them. The graves do not contain a nameless or face-less mass, but people who in the not so distant past once loved and were loved.

ACKNOWLEDGEMENTS

MY EXPERIENCE WITH this human tragedy would not come about without the tenacity and desire of International Medical Corps (IMC) in the setting up of several Ebola treatment units in West Africa to help with the outbreak. The heart-breaking scenario of grieving relatives and dying people being played out in the media nightly must have touched the hearts of those at IMC and mine in a poignantly unflagging way to motivate us to venture out to help.

Thanks to the dedicated staff of the Center for Disease Control and Prevention (CDC) training team before my deployment to Liberia for giving me the knowledge and imparting some calmness and a degree of guarded confidence in my heart for a risky mission even if it were in a simulated setting.

In a hitherto unknown perilous and desperate situation, many had risen to the occasion including the innumerable nationals and expats working and volunteering at the risk of losing their lives. There were too many of them to recount here but I would like to call out to a few individuals specifically Pero Tabby, my roommate from Kenya, for sharing our stories and supporting each other in our griefs over the loss of our patients and our joy when there were small victories. Patrick Githinji, A. Welehyou Duo and Vasco Wuokolo who worked closely with me in the ETU in Bong, their dedicated, no-nonsense, indefatigable attitude had been an inspiration. The unassuming but highly qualified Elvis Ogweno, who was responsible for setting up the ambulance services for IMC, drove out to the communities with his team almost daily to round up potentially Ebola-infected people. I was grateful to be able to witness this unsung hero and the people he trained for a couple of days performing this risky task, whereas CBS 60 minutes reporting of the Ebola Hot Zone recognized him with a mere obscure footnote.

I would also like to mention the doctors who crossed my path in Liberia: Steven Hatch for taking me in for the first time into the ETU, a very surreal time for me, Steve Whiteley, the emergency doctor who did most of the night shifts, eased me into those ungodly hours very early on in my stint in the ETU, shortly after my orientation, Colins Buck, and Trish Henwood.

In Sierra Leone, I would like to acknowledge the doctors and nurses working with me: Jon Yoder, Jean Andersson-Swayze, Kashif Islam, Brima M. Sesay, Jattu Navo, Gilbert Lekakeny, Celine Abande, Paige Fox, and many others.

The supporting casts of the ETU and ETC of both Liberia and Sierra Leone, the WASH team, the donning and doffing teams, particularly Sophie Bellorh Jarpah and Augustin Mulbah at Bong whose meticulous care in the dressing and decontamination process testified to their forbearing patience in the insufferable heat and humidity of West Africa. There were many more whose names I do not know, to all of them I give my heartfelt thanks.

To many of the patients and their relatives who,

despite their dire conditions, allowed me to take the painful journey with them, I owe them an enormous debt of gratitude. They had shown me their courage, resilience, and fortitude in the face of unspeakable horrific tragedy. They taught me humility and showed me how blessed I was.

I dedicate this book to all who were and are still affected by the Ebola virus and the Ebola fighters who put their lives on the line.

Thanks also to my literary agent, Jennifer Lyons, who helped me find a home for my first book and for her advice for a first-time author, for my editors Paul deAngelis and Hannah Bennet for their insights and patience, Sara Brady for her careful editing, for Meghan Kilduff and Allyson Fields of the production department for their careful work in making this book as beautiful as it could be.

Finally, to my family, my husband, Scott and my three children, Tim, Cara, and Charles who put up with my going away for long stretches of time to places they hear about in the news, to remote, unchartered territories harboring unknown and unforeseen dangers, I am grateful for their unconditional love and support and for bearing the burden of praying that in the end everything would turn out well.